Events Management

Events Management

Editor

Leo Rask

scitus
academics

Events Management
Edited by **Leo Rask**

ISBN: 978-1-68117-426-6
Library of Congress Control Number: 2016937579

SCITUS Academics LLC,
www.scitusacademics.com
Box No. 4766, 616 Corporate Way,
Suite 2, Valley Cottage,
NY 10989

Printed in United States of America on Acid Free Paper

Preface

Event management is the coordination, operation and planning of all people, teams and functions to create all types of events. Management of projects is used to create and develop large-scale events such as festivals, conferences, ceremonies, formal parties, concerts or meetings. It involves studying the brand, identifying the target audience, developing the concept of the event, planning logistics and coordinating technical aspects before the event begins. The planning and coordination process of the event is normally referred to as event planning and may include budgeting, scheduling, location selection, coordination of transport and parking, arrangement for entertainers, decoration, event security, catering and emergency planning. Event managers are needed in almost every business—entertainment, finance, government, retail, fashion, sport, music, so that a wide range of exciting challenges and unique circumstances can meet their day-to-day work. Events normally require a wide range of teams, often including several external suppliers. Successful events depend on a high standard of work, flexibility and commitment to good cross team communication for each team member. The event industry is all about exceeding all parties expectations and design is one of the most important ways of impressing. Event managers must be genuine outside thinkers who have a dynamic approach to problem solving. Event management is a fast, diverse and engaging industry. Even in times of economic recession, events remain invaluable to companies as a sure way to build brand reputation, build consumer confidence and recall experiences.

The Book, Events Management is a consistent, planned and targeted approach to a variety of activities.

Table of Contents

CHAPTER 1

Absolute price as a determinant of perceived service quality in hotels: a qualitative analysis of online customer reviews

Michail N. Giannakos[1], Ilias O. Pappas[2], Patrick Mikalef[2]

[1]Department of Computer and Information Science, Norwegian University of Science and Technology (NTNU), Sem Saelands vei 7-9, 7491 Trondheim, Norway

[2]Department of Informatics, Ionian University, Plateia Tsirigoti 7, 49100 Corfu, Greece

ABSTRACT

User generated content and especially customer-generated reviews are becoming a prominent information source for travellers making hotel booking decisions. Building upon the dimensions of the SERVQUAL model, the aim of this study is to identify why hotel customers have disparate opinions regarding perceived service quality during their stays. In order to do so, we distinguish between high and low absolute priced hotels. We adopt this categorisation based on prior studies, which identify price as a strong determinant of hotel customers' perceptions and decisions. By applying qualitative data analysis methods on customer reviews from one of the world's leading online hotel reservation agencies (Booking.com), we show that absolute price has an impact on how customers perceive various dimensions of service quality. In particular, the number of positive reviews is significantly different for low and high priced hotels when examining the dimensions of*tangibility* and *empathy*. Additionally, our research reveals that the number of negative reviews also differentiates significantly for low and high priced hotels, not only for the dimensions of *tangibility* and *empathy*but also with regard to *responsiveness* and *reliability*. Based on these findings we highlight implications for researchers and practitioners and suggest future directions for research.

KEYWORDS

user generated content, online hotel booking, price, SERVQUAL, convenience, perceived service quality

1 INTRODUCTION

Consumers use a variety of information sources to infer the quality of goods and services. To this extent the internet has revolutionised the way hotel customers search information and compare facilities and accommodation before booking a hotel. However, identifying hotels that meet customer requirements has become a burden, primarily due to information overload on the web and because of distrust towards travel agent websites, which are often misleading. Due to these shortcomings consumers have turned to a more reliable source of information about hotel service quality, which is based on reviews by past visitors, provided through online booking systems

(Bajari and Hortacsu, 2004; Ye et al., 2009). Information of this nature is termed user-generated content (UGC) and is growing at an increasing rate in both quantity and importance for the online hotel market (Ye et al., 2011). This phenomenon can be largely attributed to the rapid growth of Web 2.0 applications that facilitate UGC (Sigala, 2009).

Previous research has argued that absolute price has an impact on the way customers will express their satisfaction (Carvell and Quan, 2008; Yacouel and Fleischer, 2012). However, the impact of price as a determinant of expected service (ES) quality still remains unclear. To date the majority of studies have examined how reference price, as an aspect of a product or a service purchased, has an influence on the perceived value realised (Dodds et al., 1991). The rationale of these studies is based on the logic that a customer will relate the value of a product/service to its price. This association is also acknowledged as value-for-money (Sweeney and Soutar, 2001) and examines perceptions of quality under the prism of relative price paid. Therefore, the satisfaction that customers will express is dependent upon the relative price that they have paid (Bojanic, 1996). Although these studies may provide meaningful information regarding pricing schemes for products/services, they do not assess how customer expectations may vary based on an absolute price, and how these expectations impact their overall satisfaction. This is mainly because they do not examine the relationship between expectations and confirmation, and factors that shape these expectations. The expected value of a product or service is noted to be very different from the realised satisfaction (Sweeney and Soutar, 2001). This study aims to fill this gap by examining if the absolute price listing for a hotel has an impact on their expectations about service quality and in sequence their satisfaction, measured through positive of negative word-of-mouth. Our theoretical reasoning is grounded on the idea proposed in expectation-confirmation theory, which posits that consumers develop expectations; if these expectations are confirmed this will lead to satisfaction (Oliver,

1980). We suggest a variation of this argumentation, and postulate that expectations are influenced by absolute price, therefore satisfaction (expressed as positive or negative reviews) will be dependent upon the discrepancy between expected and realised service quality attributes. Consequently, we infer that satisfaction is indirectly influenced by the absolute price of a hotel.

In order to operationalise our research, an empirical study was conducted using data extracted from one of the world's leading online hotel reservation agencies. The dataset collected was comprised of customer-generated reviews retrieved from the online travel agency Booking.com (Priceline, NASDAQ: PCLN). Satisfaction was operationalised through the SERVQUAL (Parasuraman et al., 1985) framework as perceived service (PS) quality with the addition of convenience (Akbaba, 2006). Findings are expected to shed light on how to increase effectiveness of marketing strategies when promoting expensive and inexpensive hotels respectively.

To this end, the present study aims to identify those service quality dimensions that are of greatest importance for customers. Additionally, we examine how service quality perceptions differ for customers that select low and high absolute priced hotels. Therefore, the specific objectives of the study are as follows:

1 to identify which dimensions of service quality are more important (more frequently reported) from a customers' perspective in the hotel industry

2 to examine potential differences in each service quality dimension among customers that pay for expensive and inexpensive hotels.

The paper is structured as follows. In the next section, the theories underlying the key variables studied in the paper are discussed and our research hypotheses are presented. Section 3 presents the methodology as well as the procedures adopted for gathering and coding data from customer reviews. Section 4 presents empirical results, while the last section of the paper raises the key theoretical and practical implications of this research and discusses several ideas on further research in the area.

2 THEORETICAL BACKGROUND AND RESEARCH HYPOTHESES

2.1 Service quality and online travel reviews

Service quality refers to the ability to fulfil customers' requirements; both stated and implied (Lin and Su, 2003). Regarding the hotel industry, PS quality refers to how well customer expectations are met by the hotelier. The degree to which service quality criteria are met is considered as an important determinant of hotel customers' satisfaction and acceptance (Christou, 2006). Hence, service quality is regarded as a highly important aspect and has been identified as an enabler of a competitive advantage in the hotel industry (Yang and Yang, 2011).

However, measuring the quality of a service is a complex task as opposed to when measuring the quality of goods. This is primarily due to the multi-dimensional and subjective nature of services. Despite this fact, a number of proposed frameworks have been put forth, with one of the most popular and successful assessment tools of service quality being SERVQUAL, originally proposed by Parasuraman et al. (1985). While the SERVQUAL framework was originally measured by employing ten dimensions of service quality, a later version by Parasuraman et al. (1988) merged them into five. The five dimensions of the later framework include *Tangibility* (TANG), *Reliability* (REL), *Responsiveness* (RES), *A ssurance* (ASS), and *Empathy* (EMP). The definitions for each of the dimensions are listed below (Parasuraman et al., 1988):

1 *Tangibility* includes the physical evidence of services, such as physical facilities, tools or equipment, appearance of employees, and other customers.

2 *Reliability* involves the consistency of performance and dependability. It means that the firm performs the service correctly for the first time and honours its promise.

3 *Responsiveness* concerns the willingness or readiness of employees to provide service. It involves timeliness of services (e.g., setting up an appointment quickly).

4 *Assurance* refers to the knowledge, courtesy of employers and their ability to convey trust and confidence.

5 *Empathy* means providing attention to every individual customers demands by the staff.

Based on suggestions from literature we incorporate the dimension of *Convenience* as an additional aspect of service quality (Kumar et al., 2010). *Convenience* has been noted as being of paramount importance for hotel customers, to such an extent that it could possibly affect their overall evaluation of service quality (Akbaba, 2006; Berry et al., 2006). This dimension of service quality emphasises on hotel location, and the ease and adequacy of offered services, an aspect especially important to customers with a busy agenda.

6 *Convenience* means accessibility and ease for the customers.

Therefore, this study is grounded on the extended SERVQUAL framework, which is comprised of the six aforementioned dimensions.

Service quality is often conceptualised as the comparison of service expectations with actual service perceptions (Pizam and Ellis, 1999). According to a study by Pizam and Ellis (1999), service quality leads to the attraction of new customers and leads to an overall increase of business performance. Consequently, the reliability of information provided to consumers is critical, since it has a direct impact on the expected level of service quality and their reaction to the actual service provided.

Information gathering by consumers is performed both online and offline, however, online transactions are rapidly gaining momentum and replacing traditional methods (Buhalis and Licata, 2002). The information gathering procedure has been greatly improved by the appearance of online travel agencies, which provide consumers with an effective way of gathering information regarding hotels and their facilities. Through such online travel agencies (e.g., Booking.com, Expedia.com) consumers can browse through a list of hotels at a specific destination, find and compare hotel information (e.g., prices, star rating, location, facilities) and book the one of their preference. In the evaluation and selection of the hotel to be booked, consumers rely at a growing rate on comments made by past customers. Such online travel reviews are defined as user generated content that is usually posted on hotel, company or travel agent websites (Ye et al., 2009). The reviews posted are in most cases rich in information about service quality aspects, post experience evaluation as well as overall evaluation. Gretzel and Yoo (2008) found that almost all potential travellers find reviews to be helpful when making a decision, by increasing the likelihood of congruency between expectations and actual purchases. Also, based on reviews customers may be able to get new ideas or find alternative ones. Online travel reviews are advantageous in examining customers' perceptions, and compared to other methods such as questionnaires or interviews, they are considered as a compelling and more reliable method to investigate those perceptions (Ye et al., 2012).

The SERVQUAL framework has been employed as the theoretical basis in many empirical studies for measuring customers' PS quality (Devaraj et al., 2002). The principal advantage of SERVQUAL is that it has been proven to be applicable across a large range of service contexts. Parasuraman et al. (1991) indicate that SERVQUAL could undergo minor modifications in order to be valid in almost any context. To date, SERVQUAL has been applied in various settings of the hotel industry; including consumer opinions, IT services, and healthcare areas (Albacete-Sáez et al., 2007). The methodology by which the SERVQUAL instrument has been operationalised is primarily through surveys. In the present study, the dimensions that comprise the SERVQUAL frameworks are extracted by using novel approaches on customer electronic reviews. These reviews were obtained from one of the leading travel agency websites on which content analysis techniques were applied in order to classify the content into one or more dimensions of the modified SERVQUAL instrument. The prime differentiator with past studies is that the approach followed in this study centres on those dimensions of service quality which consumers regard as most important, i.e., those that are mentioned in their reviews.

2.2 Pricing in hospitality industry and research hypotheses

In their attempt to deal with *TheLemon Problem* (Akerlof, 1970) within the hotel industry, professional certifiers were developed in order to rate hotel facilities (five-star rating systems). The main idea of these certifiers is to provide potential customers with a quantified measure of the service and facilities to be expected by a hotel.

However, these rating systems mainly reflect physical attributes and not customers' attitudes, they are not always reliable, and each country may base their scaling on different standards (Yacouel and Fleischer, 2012). Before the emergence of the internet, travel agents and tour operators served as intermediaries between hoteliers and prospective guests, helping with the search and selection process, thus mitigating the information asymmetry problem. Clerides et al. (2008) examined the role of travel agencies as reliable quality assessors in the hotel market as they directly connected with the market and found that the travel agencies' rating of hotels was better matched with prices than with the national five-star system. Although pricing methods refer to the specific formulas used in order to set a price and include several aspects (e.g., discounts, coupons), pricing policies are associated with the actual prices (Lowengart and Mizrahi, 2000). The main underlying goal of the above studies is to develop a better fit between quality criteria and pricing, thus mitigating the problem of information asymmetry. The main premise of these investigations is that price should be appropriated in such a way in order to provide hotel customers of an indication of the quality attributes that should be expected. Since relative price is a measure that rates price towards service quality perceived, and requires that consumers have utilised the product service, it a measure that we deem as unfit to impact on consumers' expectations regarding service quality. Hence, we employ the measure of absolute price to determine how it may impact consumers' expectations. Since the discrepancy between expectations and actual service will determine the PS quality we advocate that absolute price is an important determinant of overall consumer satisfaction.

Cunningham et al. (2002) reported significant differences in service quality perceptions among customers with different characteristics. Differentiated pricing involves offering different prices to customers on the basis of numerous criteria. In general, customers tend to form higher quality perceptions and give higher rating to hotels, when the price they paid for a room seems to be cheaper than it should be (Oh, 2003). This idea is grounded on the price measure as a relative price, since what is considered as cheap and expensive is subjective. In this research we investigate how PS quality differentiates on the basis of actual and not perceived price. Our prime target is to identify if expensive and inexpensive hotels impact consumers perceptions about the quality of service offered to them. Ye et al. (2012) found that when customers evaluate a hotel, the actual price paid positively affects their PS quality. This implies that the service quality expectations increase in accordance with the paid price. This association is more significant when examining expensive hotels, with the negative association between inexpensive hotels and PS quality being dampened.

In their study, Parasuraman et al. (1985) advocate that realised service quality depends on the discrepancy between ES and PS. Ladhari (2009) define behavioural intentions as a factor influenced by willingness to pay more money and mention that favourable behavioural intentions include paying a premium price. They found that these intentions are positively affected by PS quality. Consequently, it is likely that a relation exists between the willingness to pay more and PS quality and its dimensions, although such relation is not examined on Ladhari (2009). This disparity is evident in the case of expensive hotels from which consumers expect a high level of service quality (Ye et al., 2012). This association of price being a determinant of service quality is also noted in literature (Bitner and Hubbert, 1994). Hence, the quality of service expected by customers relies on the price that they have paid (Briggs et al., 2007). Extending on this notion, it can be inferred that low absolute priced hotels are not expected to offer high quality services; therefore, customers will have lower expectations. This means that the effort required from hoteliers to cover these requirements is significantly less in comparison with expensive hotels, leading in sequence to higher levels of PS quality by customers.

Numerous studies have examined the idea of service quality at the level of the SERVQUAL dimensions, looking at them in isolation. One of these studies conducted by Hartman (1989) found that business travellers exhibit a willingness to pay for amenities such as room service, availability of parking, and guaranteed reservations in a luxury hotel, linking as such price with service quality. Monty and Skidmore (2003) extended findings in this matter and identified facilities such as a hot tub, a private bath and a larger room as having an impact on the price paid. Based on the above findings we can assume that hotel customers that pay a high absolute price will have high expectations regarding the *tangibility* dimension of service quality. Collins and Parsa (2006) identified many factors that influence pricing decisions, including location, size and amenities, which place emphasis on the importance of *tangibility* as well as that of *convenience*. A study by Briggs et al. (2007) found that the main reason for low PS quality was related to the absence of a personal touch and the way that which the hotel dealt with complaints, thus highlighting the importance of *empathy* and *assurance*. Zhang et al. (2011) found that the first requirement for expensive hotel customers is location, which is an element of *convenience*, while *tangibility* factors (e.g., room design, facilities) are the most important ones for customers of absolute low priced hotels. The aforementioned studies link hotel facilities and service quality with a price, thus showing that customers will pay more for additional services. Reversing this logic and supported on the findings on Webster (1991), which notes that customers are affected by past experiences, we postulate that customers who pay for expensive hotels have high expectations of service quality. Since positive feedback (an outcome of satisfaction) is a result of expectations being surmounted and desired service quality attributes being fulfilled, we advocate that in the case of expensive hotels the margin to do so is limited in comparison to inexpensive hotels (Zeithaml and Berry, 1993). Therefore, positive feedback regarding the PS quality dimensions for customers of expensive hotels will be significantly less than those of inexpensive hotels. Following the same train of thought, we postulate that expectations of customers of low absolute priced hotels can easily be met to the adequacy level. Hence, it is more likely that the number of negative comments by hotel customers of this category will be significantly less than those of high absolute priced hotels.

In our study, we measure PS quality using dimensions adopted from previous studies (Chang, 2009; Fernández and Bedia, 2005; Juwaheer, 2004; Voss, 2003). Based on the above argumentation we develop the following hypotheses:

H1a High paying hotel customers have significantly fewer positive reviews regarding empathy than low paying hotel customers.

H1b High paying hotel customers have significantly more negative reviews regarding empathy than low paying hotel customers.

H2a High paying hotel customers have significantly fewer positive reviews regarding convenience than low paying hotel customers.

H2b High paying hotel customers have significantly more negative reviews regarding convenience than low paying hotel customers.

H3a High paying hotel customers have significantly fewer positive reviews regarding reliability than low paying hotel customers.

H3b High paying hotel customers have significantly more negative reviews regarding reliability than low paying hotel customers.

H4a High paying hotel customers have significantly fewer positive reviews regarding assurance than low paying hotel customers.

H4b High paying hotel customers have significantly more negative reviews regarding assurance than low paying hotel customers.

H5a High paying hotel customers have significantly fewer positive reviews regarding responsiveness than low paying hotel customers.

H5b High paying hotel customers have significantly more negative reviews regarding responsiveness than low paying hotel customers.

The only exception to this rationale is the service quality dimension of *tangibility*, because hotel customers exhibit high willingness to pay for such attributes (Hartman, 1989) and this dimension is much more objective than the previous five and the only one that is intangible. According to the study by Johnston (1995), tangible aspects of service quality are usually perceived as less important and it is therefore less likely that customers will make negative reviews. This applies especially for expensive hotels, in which hoteliers place much effort (interior design, furniture, etc.). However, in inexpensive hotels, which usually manage to offer low prices due to the absence of tangible artefacts, we assume that customers will not focus on posting positive comments. As such, we propose that there is a significant positive difference on the disparity between ES and PS, among the high and low paying hotel customers; hence the following hypotheses are formulated:

H6a High paying hotel customers have significantly more positive reviews regarding *tangibility* than low paying hotel customers.

H6b High paying hotel customers have significantly fewer negative reviews regarding *tangibility* than low paying hotel customers.

3 RESEARCH METHODOLOGY

3.1 Data collection

The data used in this study were retrieved from Booking.com (Priceline, NASDAQ: PCLN), one of the world's leading online hotel reservations agencies. Booking.com allows past hotel guests who have booked a hotel room through their website to write reviews. Consequently, there are no unreliable posts by users that have not stayed at the hotel... In addition, Booking.com also monitors the hotel descriptions published on its website, ensuring that the hotels' actual informative data match those that are published.

For the data collection procedure (Figure 1), ten large cities and popular holiday destinations were randomly selected from the USNews1 and then ten customer reviews (including positive and negative as provided by Booking.com based on customers' categorisation) of ten hotels (five expensive and five inexpensive based on Booking.com price ranks) of each city were selected.

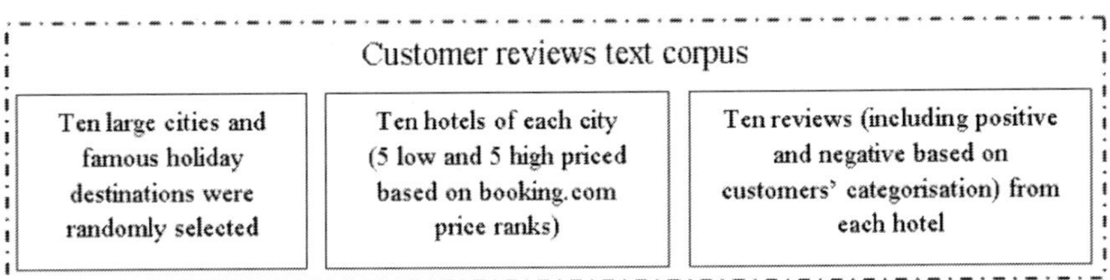

Figure 1 Data collection procedures

These reviews covered a text corpus of approximately 120 pages. Roughly half of the corpus represents expensive hotels and the rest concerns inexpensive ones. In addition, the corpus almost equally represents positive and negative comments. By following these data collection procedures (Figure 1), we achieved to have four equally representative categories. As such, from the total of 120 pages, four categories (positive-inexpensive, positive-expensive, negative-inexpensive and negative-expensive) with almost the same length (30 pages) were identified.

3.2 Coding procedures and protocol

After completing the customer reviews collection (text corpus), we proceeded in performing a qualitative text analysis of the case studies. The methodology of Mayring (2000) was selected, which combines several techniques for text analysis into a systematic process. According to Mayring (2000), the category system or coding categories might derive either from an existing theory (deductive category application), or develop during the analysis from the text corpus (inductive category development). In our case we used SERVQUAL as our theoretical basis and in conjunction with literature suggestions (Akbaba, 2006; Kumar et al., 2010) decided to include the dimension of *convenience*. For the coding procedure and for the evaluation of the coding results (Figure 2), we employed the software package MaxQDA (http://www.maxqda.com).

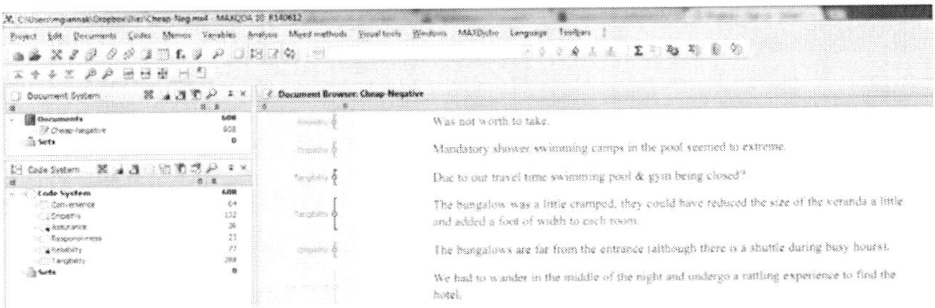

(a) (b)

Figure 2 Coding procedure through MaxQDA, (a) the code categories 'system' (b) and the text corpus (see online version for colours)

In particular for the coding, two researchers with experience on qualitative analysis and the SERVQUAL framework coded the text corpus (customer reviews) independently. In order to facilitate the work of the two researchers prior to the coding we developed and assisted them with the following coding protocol (Table 1), which contains the operational definitions of dimensions followed by an example for each dimension separately as identified in the literature.

4 RESEARCH FINDINGS

After the coding procedures, 3,693 codes were identified. A Cohen kappa test was conducted to examine the inter-reliability of the coding. The number of codes and the index of reliability at each category are presented in Table 2.

Table 1 Coding protocol

Dimensions	*Definition*	*Example*
Empathy (EMP)	The provision of caring, individualised attention the firm provides its customers.	The staff acknowledges customers at the reception with a verbal greeting.
Convenience (CONV)	Convenience involves comfort, accessibility and easiness to the customers.	Hotel locates in the great area (near centre, train station, bus station, airport, etc.).
Reliability (REL)	The ability to perform the promised services dependably and accurately.	The staff performs tasks that have been promised to guests and resolves any problem encountered.
Assurance (ASS)	The knowledge and courtesy of employees and their ability to convey trust and confidence.	The staff can handle effectively complaints and problems from the guests.
Responsiveness (RES)	The willingness to help customers and provide prompt service.	The customers are offered alternative accommodation at a nearby hotel. Reservation cards are ready to sign upon arrival.
Tangibility (TANG)	The appearance of physical facilities, equipment, personnel and communication materials.	The front office staff is well dressed.

Table 2 Index reliability

Review category	*Hotel category*	*Codes*	*Cohen kappa*
Positive	Inexpensive	1,083	0.82
Positive	Expensive	1,028	0.78
Negative	Inexpensive	723	0.70
Negative	Expensive	859	0.74

Every category surpassed the recommended threshold (Cohen kappa = 0.70) (Lombard et al., 2010), indicating high inter-coder reliability. Table 3 exhibits the exact results of the coding procedure.

Table 3 Coding results

		Rater A											
		Positive						Negative					
		EMP	*CONV*	*REL*	*ASS*	*RES*	*TANG*	*EMP*	*CONV*	*REL*	*ASS*	*RES*	*TANG*
Rater B, Inexpensive	EMP	*171*	2	4	8	5	9	*112*	0	6	7	4	13
	CONV	2	*232*	0	9	1	6	3	*59*	0	9	2	9
	REL	9	7	*100*	5	7	3	8	5	*69*	2	8	11
	ASS	2	7	3	*121*	8	4	4	2	5	*31*	7	5
	RES	4	2	6	11	*61*	5	2	5	3	8	*23*	6
	TANG	8	8	5	0	7	*241*	15	3	8	4	0	*265*
Rater B, Expensive	EMP	*121*	8	5	6	2	7	*179*	9	3	8	3	5
	CONV	4	*178*	1	12	8	8	14	*51*	1	11	9	12
	REL	9	3	*107*	8	4	1	10	7	*122*	1	3	3
	ASS	7	13	5	*102*	4	4	5	8	7	*42*	7	3
	RES	2	0	4	9	*67*	5	0	3	4	6	*47*	7
	TANG	16	11	3	9	6	*269*	9	6	5	0	5	*244*

Note: RES: responsiveness; ASS: assurance; CONV: convenience; TANG: tangibility; EMP: empathy; REL: reliability.

From the total of 3,693 codes we have an agreement on 3,014 codes. Based on the agreement codes, we identify the following service dimensions, which are presented in descending order (Table 4).

Table 4 Frequency appearance priority of service dimensions based on the number of customer reviews

Dimensions prioritised by the frequency they reported	*No. of codes (%)*
Tangibility (TANG)	1,019 (33.81%)
Empathy (EMP)	583 (19.34%)
Convenience (CONV)	520 (17.25%)
Reliability (REL)	398 (13.21%)
Assurance (ASS)	296 (9.82%)
Responsiveness (RES)	198 (6.57%)

From the total agreement codes, 1,770 (58.73%) are positive and 1,244 (41.27%) are negative, 1,485 (49.27%) are generated for inexpensive and 1,529 (50.73%) for expensive hotels. In particular, 926 (30.72%) of the codes are positive and generated from customers of low absolute priced hotels, 844 (28.00%) of the codes are positive and generated from customers of high absolute priced hotels, 559 (18.55%) of the codes are negative and generated from customers of low absolute priced hotels and 685 (22.73%) of the codes are negative and generated from customers of high absolute priced hotels. The descriptive statistics in Table 4present the coding frequency of each dimension. To examine the differences on the number of reviews in each quality dimension amongst customers of high and low priced hotels, a Pearson's chi-square test was conducted. The results of chi-square test are summarised in the following table (Table 5).

Table 5 Descriptive statistics and Pearson's chi-square test results between customers of expensive and inexpensive hotels

Factors	*Positive*						*Negative*					
	Codes (%)		*Sign.*	*Chi*	*df*	*Result*	*Codes (%)*		*Sign.*	*Chi*	*df*	*Result*
	Inexpensive	*Expensive*					*Inexpensive*	*Expensive*				
EMP	171 (5.67%)	121 (4.01%)	*0.048**	*3.93*	1	H1a (A)	112 (3.72%)	179 (5.94%)	*0.046**	*3.98*	1	H1b (A)
CONV	232 (7.70%)	178 (5.91%)	0.119	2.44	1	H2a (R)	59 (1.96%)	51 (1.69%)	0.079	*3.08*	1	H2b (R)
REL	100 (3.32%)	107 (3.55%)	0.275	1.19	1	H3a (R)	69 (2.29%)	122 (4.05%)	*0.022**	*5.22*	1	H3b (A)
ASS	121 (4.01%)	102 (3.38%)	0.584	0.30	1	H4a (R)	31 (1.03%)	42 (1.39%)	0.68	*0.17*	1	H4b (R)
RES	61 (2.02%)	67 (2.22%)	0.308	1.04	1	H5a (R)	23 (0.76%)	47 (1.56%)	*0.048**	*3.92*	1	H5b (A)
TANG	241 (8.00%)	269 (8.93%)	*0.044**	*4.06*	1	H6a (A)	265 (8.79%)	244 (8.10%)	*0.007***	*7.37*	1	H6b (A)
Total	926 (30.72%)	844 (28.00%)					559 (18.55%)	685 (22.73%)	3014 (100%)			

Notes: Significance level, $**p < .01$; $*p < .05$. EMP: empathy; CONV: convenience; REL: reliability; ASS assurance; RES: responsiveness; TANG: Tangibility; (A): accepted; (R): rejected.

As Table 5 shows, in the case of positive reviews, the price paid exhibits significant effect on the *tangibility* and*empathy* dimensions. Observing the number of codes, customers make significantly more positive comments regarding the *tangibility* dimension of expensive hotels. On the other hand, customers of high absolute priced hotels make significantly fewer positive comments regarding the *empathy* dimension. For the case of negative reviews, the price paid exhibits significant impact on the *responsiveness, tangibility, empathy* and *reliability*dimensions. Observing the number of codes, customers of high absolute priced hotels make significantly more negative comments regarding the *responsiveness, empathy* and *reliability* dimensions. On the other hand, customers of low absolute priced hotels make significantly more negative comments regarding the *tangibility*dimension.

5 DISCUSSION AND CONCLUSIONS

We have examined how the absolute price of hotels impacts the perceived quality of services realised by customers. The attribute of quality is examined through a novel approach by using customer electronic reviews (word-of-mouth) posted on a travel agency website. We reverse the directionality of hypotheses of Zhang et al. (2011) who examine what hotel attributes lead customers an increased willingness to pay, and find that the price paid by hotel customers has a significant impact on how they perceive different elements of service. This is done by codifying online customer reviews according to the dimensions of SERVQUAL with the addition of*convenience.* Grounded on past studies regarding cost-quality associations, we develop a set of hypotheses (H1a–H6a and H1b–H6b) which are then analysed. Findings reveal that several dimensions of PS quality are significantly differentiated based on the absolute pricing of the hotel. In particular, the dimensions of*responsiveness, tangibility, empathy* and *reliability* exhibit such a differentiation.

As the majority of the reviews address it, *tangibility* is the most important dimension (more frequently reported) from the customers' perspective (Table 4). Through the analysis of this specific dimension, we find that customers of high absolute priced hotels make significantly more positive comments than those of inexpensive hotels (H6a accepted) and customers of low absolute priced hotels make significantly more negative reviews than those of high absolute priced hotels (H6b accepted). However, this may be based on the fact that expensive hotels have more facilities and as a result customers of expensive hotels make more positive and fewer negative comments. Ergo, it seems that the hotel price reduction does not affect customers' perceptions regarding the lack of some facilities.

For the dimensions of *assurance* and *convenience* there is no significant difference between the number of reviews (positive and negative) for customers of expensive and inexpensive hotels (H2a,b and H4a,b are rejected). In addition, observing Figure 3 we can infer that customers' perceived *convenience* and *assurance*are important supplementary services of a hotel (due to the many positive reviews) and not important requirements (due to the few negative reviews in Figure 4). On the one hand regarding *assurance*, our results are partially consistent with previous studies that found *assurance* factors (e.g., courteous staff) to be a desirable service but not a requirement for expensive hotels (Zhang et al., 2011). On the other hand with regard to convenience, our results partially contradict Zhang et al. (2011), who found it to be the most important requirement for high paying customers. For the case of *empathy*, there is a significant difference between the number of reviews (both positive and negative) customers of expensive and inexpensive hotels (H1a and H1b are accepted). These results verify that customers of expensive hotels have high expectations for the *empathy* dimension. The dimension of *reliability* has great similarity in the number of reviews in each category (Figures 1 and 2). Regarding the positive reviews, there is no significant difference between customers of low and high absolute priced hotels (H3a is rejected). For negative reviews however, there is a significant difference since customers of expensive hotels have made significantly more negative comments. Therefore, H3b is accepted and verifies the difference in perception of service quality for customers of expensive and inexpensive hotels.

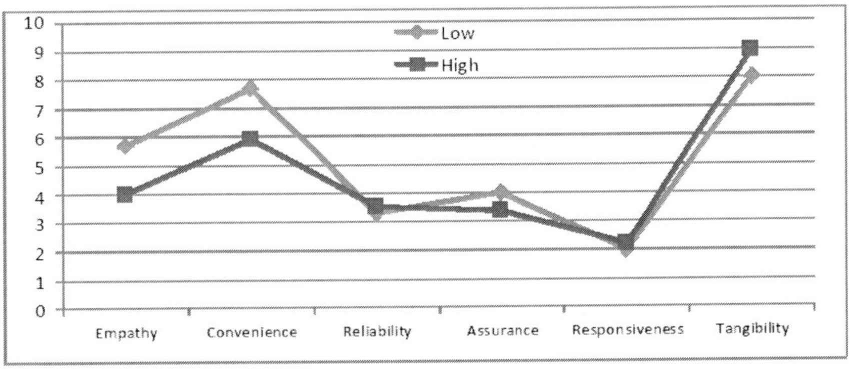

Figure 3 Frequency of positive reviews from customers of low and high absolute priced hotels on every dimension (see online version for colours)

As for the *assurance* dimension, there is no significant difference between the number of positive and negative reviews (H4a and H4b are rejected). *Responsiveness* involves the prompt and proactive response towards customer needs. The results on this dimension indicate an insignificant difference on the number of positive reviews (H5a rejected) and a significant difference on the number of negative reviews (H5b accepted). Hence, customers of expensive hotels have made significantly more negative reviews with regard to this dimension. As a result *responsiveness* is not regarded as an optional service (due to H5a rejection) but as prerequisite (due to H4b acceptance).

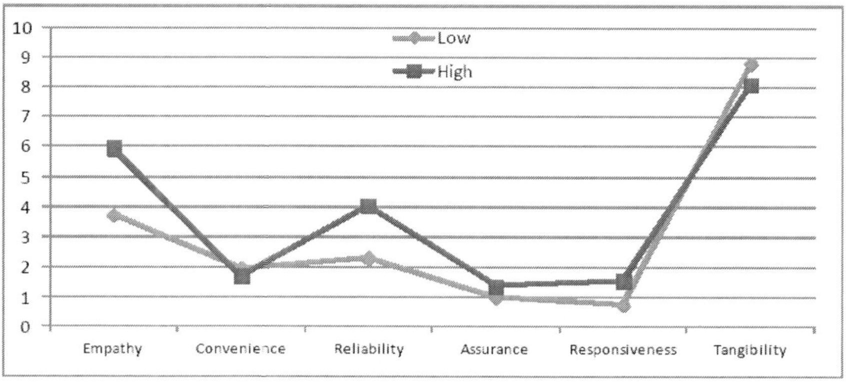

Figure 4 Frequency of negative reviews from customers of low and high absolute priced hotels on every dimension (see online version for colours)

5.1 Implications

The main theoretical implication of this research relates to filling in the knowledge gap of the impact of absolute price paid on the prime determinants of visitors' PS quality based on online customer-generated reviews. This study measures the six key quality dimensions and provides meaningful results for each price and review category. In addition, the research findings reveal differences in many quality service dimensions between hotel customers of expensive and inexpensive hotels. Marketing strategies must be differentiated and focused on the appropriate quality attributes, as some dimensions appear to be more important for hotel customers. Thus, owners of inexpensive hotels should focus on increasing their *empathy*-related services, while owners of expensive hotels should focus their efforts in increasing *responsiveness, tangibility* and *reliability* of their services.

These results address a few shortcomings of previous studies on the area. Specifically, Stringam and Genders (2010) used customer online reviews to identify important commendations and concerns. Their study identified words and not dimensions; they found that cleanliness was the most important concern of the travellers, with location and comments about the staff following. Even though the identified words may be categorised to the specific dimensions of the SERVQUAL construct, not all dimensions can be included. Moreover, Stringam and Genders (2010) did not investigate the different effects of the price paid by the customers on the review they made about the hotel. Ladhari (2009) investigated service quality along with customer satisfaction and behavioural intentions. In this study, favourable behavioural intentions included paying a premium price while the opposite includes spending less money. Although the study of Ladhari (2009) included service quality factors and customers' willingness to spend more money, there was no investigation of the relation between price and PS quality.

The results of the present study highlight the differences between customer perceptions of expensive and inexpensive hotels, regarding service characteristics. This determination helps hotel managers and travel agents to have a clear idea of what expectations each customer has from hotel services, based on the amount of money he or she chooses to spend. Additionally, by identifying what dimensions are more important (frequently reported) on travel agency websites, hotel managers can focus on them without placing unnecessary emphasis on aspects, which are proven to be less important; especially for hotels focusing exclusively on each one of the customer categories. In particular, this study revealed that customers of inexpensive hotels consider *tangibility* as the most important drawback dimension on their willingness to pay for an economy hotel. To satisfy this category of customers, economy hotels should focus investments on the quality of their hotel rooms that may include room design, space and other facilities, which can be investigated by a qualitative in depth analysis of the customer reviews.

Managers of expensive hotels should primarily pay attention to physical facilities in order to increase the offered quality, since customers that choose these hotels expect modern looking equipment and appealing facilities. In addition, expensive hotels should employ well-trained staff at all positions. The staff should be able to resolve any problems that may occur promptly. Managers should invest in both physical facilities and human resources. On the other hand, managers of inexpensive hotels that do not have big amount of money to invest should focus on offering customers with caring and personalised attention in an attempt to increase their satisfaction.

5.2 Limitations and future work

While this study has used rigorous data collection and analysis procedures, some limitations need to be highlighted, which are worthy to be mentioned in the results interpretation and for consideration in future research. This study reveals the important service quality dimensions of hotels as identifying under the absolute price lens, through travel reviews posted on a travel agency website (Booking.com). However, online reviews may not provide the whole view of hotel customers' opinions. For instance, it is possible that there are customers who were delighted (or opposite) tended to write extensive reviews. Nevertheless, promotions to encourage posting online reviews by booking websites have provided the websites with better representation of customers (Ye et al., 2011). Additionally, several studies (Zhang et al., 2011; Vermeulen and Seegers, 2009) have used reviews to infer service quality. The differentiation of customers of expensive and inexpensive hotels was made based on the prices exhibited at the time of the data collection; without taking into account any potential special offer made to the customers composing the reviews. Based on this study, future research should be carried out using more service quality dimensions and investigating other parameters (i.e., hotel star rating, age, etc.). In addition, future research could enhance our understanding as to the significance of these contemporary issues by conducting quantitative survey to canvas the opinions and weighting of these issues.

The authors would like to thank A. Pateli who helped in earlier phases of this project.

AUTHOR DETAILS

Michail N. Giannakos an is a ERCIM/Marie Curie Fellow in the Department of Computer and Information Science at the Norwegian University of Science and Technology (NTNU) and a Visiting Researcher in the Center for Real-time Computing, Virginia, USA. He holds a PhD in Educational Technology and Interaction Design; he has developed and experimented with learning environments since 2008, and since that time has single- or co-authored more than 50 manuscripts published in peer-reviewed journals and conferences. He is also a program committee member on several international conferences and a Guest Editor of the *ACM Transactions on Computing Education* and the *International Journal of Emerging Technologies in Learning*. He has worked at several research projects funded by diverse sources like EC, NSF, Microsoft Research and Cheng Endowment.

Ilias O. Pappas received his BSc in Informatics (specialisation in Humanities Informatics) from the Ionian University, Corfu, Greece and an MSc degree in Informatics (specialisation in Information Systems) from the same university. He is currently working on a PhD degree at the Ionian University. His current research interests are in the areas of electronic commerce, mobile commerce, information technology adoption, customer behaviour in online environments, personalisation and recommendation agents, and customer emotions in internet marketing.

Patrick Mikalef is currently a PhD candidate in the Department of Informatics at the Ionian University, Corfu, Greece. He holds a BSc in Informatics, with a specialisation on Information Systems from the Ionian University, and an MSc in Business Informatics in the Department of Information and Computing Sciences at the Utrecht University, Utrecht, The Netherlands. His research interests include information systems management, service-oriented architectures, digital business strategies, dynamic capabilities, and social media for business.

REFERENCES

1 Akbaba, A. (2006). 'Measuring service quality in the hotel industry: a study in a business hotel in Turkey'.Hospitality Management. 25, 2, 170-192

2 Akerlof, G.A. (1970). 'The market for 'lemons': quality uncertainty and the market mechanism'. Quarterly Journal of Economics. 84, 3, 488-500

3 Albacete-Sáez, C.A. , Fuentes-Fuentes, M.M. , Lloréns-Montes, F.J. (2007). 'Service quality measurement in rural accommodation'. Annals of Tourism Research. 34, 1, 45-65

4 Bajari, P. , Hortacsu, A. (2004). 'Economic insights from internet auctions: a survey'. Journal of Economics Literature. 42, 2, 457-486

5 Berry, L.L. , Wall, E.A. , Carbone, L.P. (2006). 'Service clues and customer assessment of the service experience: lessons from marketing'. Academy of Management Perspectives. 20, 2, 43-57

6 Bitner, M.J. , Hubbert, A.R. (1994). 'Encounter satisfaction versus overall satisfaction versus quality'. Service Quality: New Directions in Theory and Practice. 72-94

7 Bojanic, D.C. (1996). 'Consumer perceptions of price, value and satisfaction in the hotel industry: an exploratory study'. Journal of Hospitality & Leisure Marketing. 4, 1, 5-22

8 Briggs, S. , Sutherland, J. , Drummond, S. (2007). 'Are hotels serving quality? An exploratory study of service quality in the Scottish hotel sector'. Tourism Management. 28, 4, 1006-1019

9 Buhalis, D. , Licata, M.C. (2002). 'The future eTourism intermediaries'. Tourism Management. 23, 3, 207-220

10 Carvell, S.A. , Quan, D.C (2008). 'Exotic reservations – low-price guarantees'. International Journal of Hospitality Management. 27, 2, 162-169

11 Chang, J. (2009). 'Taiwanese tourists' perceptions of service quality on outbound guided package tours: a qualitative examination of the SERVQUAL dimensions'. Journal of Vacation Marketing. 15, 2, 165-178

12 Christou, E. (2006). 'A qualitative analysis of consumer attitudes on adoption of online travel services'.Tourism: An International Interdisciplinary Journal. 54, 4, 323-331

13 Clerides, S. , Nearchou, P. , Pashardes, P. (2008). 'Intermediaries as quality assessors: tour operators in the travel industry'. International Journal of Industrial Organization. 26, 1, 372-392

14 Collins, M. , Parsa, H.G. (2006). 'Pricing strategies to maximize revenues in the lodging industry'.International Journal of Hospitality Management. 25, 1, 91-107

15 Cunningham, L.F. , Young, C.E. , Lee, M. (2002). 'Cross-cultural perspectives of service quality and risk in air transportation'. Journal of Air Transportation Management. 7, 1, 3-26

16 Devaraj, S. , Fan, M. , Kohli, R. (2002). 'Antecedents of B2C channel satisfaction and preference: validating e-commerce metrics'. Information Systems Research. 13, 3, 316-333

17 Dodds, W.B. , Monroe, K.B. , Grewal, D. (1991). 'Effects of price, brand, and store information on buyers' product evaluations'. Journal of Marketing Research. 28, 3, 307-319
18 Fernández, M.C.L. , Bedia, A.M.S. (2005). 'Applying SERVQUAL to diagnose hotel sector in a tourist destination'. Journal of Quality Assurance in Hospitality & Tourism. 6, 1, 9-24
19 Gretzel, U. , Yoo, K.H. , O'Connor, P. , Höpken, W. , Gretzel, U. (2008). 'Use and impact of online travel reviews'. Information and Communication Technologies in Tourism. New York:Springer Wien , 35-46
20 Hartman, R.S. (1989). 'Hedonic methods for evaluating product design and pricing strategies'. Journal of Economics and Business. 41, 3, 197-212
21 Johnston, R. (1995). 'The determinants of service quality: satisfiers and dissatisfiers'. International Journal of Service Industry Management. 6, 5, 53-71
22 Juwaheer, T.D. (2004). 'Exploring international tourists' perceptions of hotel operations by using a modified SERVQUAL approach – a case study of Mauritius'. Managing Service Quality. 14, 5, 350-364
23 Kumar, M. , Kee, F.T. , Charles, V. (2010). 'Comparative evaluation of critical factors in delivering service quality of banks'. International Journal of Quality & Reliability Management. 27, 3, 351-377
24 Ladhari, R. (2009). 'Service quality, emotional satisfaction, and behavioural intentions: a study in the hotel industry'. Managing Service Quality. 19, 3, 308-331
25 Lin, Y. , Su, H. (2003). 'Strategic analysis of customer relationship management – a field study on hotel enterprises'. Total Quality Management & Business Excellence. 14, 6, 715-731
26 Lombard, M. , Snyder-Duch, J. , Campanella Bracken, C. (2010). Practical Resources for Assessing and Reporting Inter-coder Reliability in Content Analysis Research Projects. (accessed 25

October 2012), [online]http://astro.temple.edu/~ lombard/reliability/
27 Lowengart, O., Mizrahi, S. (2000). 'Applying international reference price: market structure, information seeking and consumer welfare'. International Marketing Review. 17, 6, 525-533
28 Mayring, P. (2000). 'Qualitative content analysis'. (accessed 25 August 2012), 1, Forum: Qualitative Social Research, 2, [online] http://www.qualitative-research.net/index.php/fqs/article/view/1089/2385
29 Monty, B. , Skidmore, M. (2003). 'Hedonic pricing and willingness to pay for bed and breakfast amenities in southeast Wisconsin'. Journal of Travel Research. 42, 2, 195-199
30 Oh, H. (2003). 'Price fairness and its asymmetric affects on overall price, quality, and value judgments: the case of an upscale hotel'. Tourism Management. 24, 4, 387-399
31 Oliver, R.L. (1980). 'A cognitive model of the antecedents and consequences of satisfaction decisions'.Journal of Marketing Research. 17, 4, 460-469
32 Parasuraman, A. , Berry, L.L. , Zeithaml, V.A. (1991). 'A refinement and reassessment of the SERVQUAL scale'. Journal of Retailing. 67, 4, 420-450
33 Parasuraman, A. , Zeithaml, V.A. , Berry, L. (1985). 'A conceptual model of service quality and its implication for future research'. Journal of Marketing. 49, 4, 41-50
34 Parasuraman, A. , Zeithaml, V.A. , Berry, L. (1988). 'SERVQUAL: a multiple-item scale for measuring consumer perceptions of service quality'. Journal of Retailing. 64, 1, 12-40
35 Pizam, A., Ellis, T. (1999). 'Customer satisfaction and its measurement in hospitality enterprises'.International Journal of Contemporary Hospitality Management. 11, 7, 326-339
36 Sigala, M. (2009). 'E-service quality and Web 2.0: expanding quality models to include customer participation and inter-

customer support'. The Service Industries Journal. 29, 10, 1341-1358
37 Strinham, B.B. , Gerdes, J. (2010). 'An analysis of word-of-mouse ratings and guest comments of online hotel distribution sites'. Journal of Hospitality Marketing and Management. 19, 7, 773-796
38 Sweeney, J.C. , Soutar, G.N. (2001). 'Consumer perceived value: the development of a multiple item scale'.Journal of Retailing. 77, 2, 203-220
39 Vermeulen, I.E. , Seegers, D. (2009). 'Tried and tested: the impact of online hotel reviews on consumer consideration'. Tourism Management. 30, 1, 123-127
40 Voss, C.A. (2003). 'Rethinking paradigms of service: service in a virtual environment'. International Journal of Operations & Production Management. 23, 1, 88-105
41 Webster, C. (1991). 'Influences upon consumer expectations of services'. Journal of Services Marketing. 5, 1,5-17
42 Yacouel, N. , Fleischer, A. (2012). 'The role of cybermediaries in reputation building and price premiums in the online hotel market'. Journal of Travel Research. 51, 2, 219-226
43 Yang, C.C. , Yang, K.J. (2011). 'An integrated model of value creation based on the refined Kano's model and the blue ocean strategy'. Total Quality Management & Business Excellence. 22, 9, 925-940
44 Ye, Q. , Law, R. , Gu, B. (2009). 'The impact of online user reviews on hotel room sales'. International Journal of Hospitality Management. 28, 1, 180-182
45 Ye, Q. , Law, R. , Gu, B. , Chen, W. (2011). 'The influence of user-generated content on traveler behavior: an empirical investigation on the effects of e-word-of-mouth to hotel online bookings'. Computers in Human Behavior. 27, 2, 634-639
46 Ye, Q. , Li, H. , Wang, Z. , Law, R. (2012). 'The influence of hotel price on perceived service quality and value in e-tourism: an

empirical investigation based on online traveler reviews'. Journal of Hospitality and Tourism Research.

47 Zeithaml, V.A. , Berry, L.L. (1993). 'The nature and determinants of customer expectations of service'. Journal of the Academy of Marketing Science. 21, 1, 1-12

48 Zhang, Z. , Ye, Q. , Law, R. (2011). 'Determinants of hotel room price: An exploration of travelers' hierarchy of accommodation needs'. International Journal of Contemporary Hospitality Management. 23, 7, 972-981

CHAPTER 2

The Social Impacts Of A Special Event On The Host City: A Conceptual Framework And A Case Study Of The Cincinnati Flying Pig Marathon

Julie Cencula Olberding[1], Douglas J. Olberding2

[1]Northern Kentucky University,

424 Founders Hall, Highland Heights, KY 41099, USA

[2] Xavier University, Joseph Building, 3rd Floor, Cincinnati, OH 45207, USA

ABSTRACT

This article provides a conceptual framework to study the potential impacts of a special event on the host city's social assets. This framework is more encompassing than traditional studies of social impact, which tend to focus only on city image; it includes local pride, social networks, and support of social causes. It then presents a case study of the Cincinnati Flying Pig Marathon and its social impacts on the host city, using unique data from surveys of participants (2002 and 2008) and volunteers (2012). The analysis finds some evidence that this event has had positive effects on Cincinnati's social assets.

KEYWORD

special events, host city, social impacts, city image, local pride, social networks, social causes, participants, volunteers

1 INTRODUCTION

Foot races, festivals and other special events are phenomena that have been around almost since the creation of the cities that host them. But interest in and studies of how these events impact host cities have come to the surface in relatively recent years. In particular, a number of studies have analysed the economic impact of marathons, which are 26.2-mile foot races, on the host cities. Some of these studies have found that recent major marathons have had a huge economic impact on the host city, such as $138 million for the 2012 Boston Marathon (Johnston, 2012) and $340 million for the 2011 New York City Marathon (Belson, 2011). However, most of these analyses have not been subject to scholarly review, and critics have asserted that they overestimate economic impacts (e.g., Porter, 1999; Crompton, 1995). Only one economic impact analysis of a marathon seems to have survived the peer review process and been published in an academic journal (Cobb and Olberding, 2007). The authors develop a method to correctly account for the impact of local marathon participants, which results in a more conservative measure of economic impact than previous analyses. Applying this method, they find that the economic impact of the Cincinnati Flying Pig Marathon is about $9 million to $10 million per year (Cobb and Olberding, 2007, 2008, 2012).

A few scholarly articles have looked beyond the economic impacts of special events to the social impacts on the cities or communities that host them. These studies tend to focus on one social asset: city image. This article develops a more comprehensive conceptual framework of a community's social assets – including not only city image but also local pride, social networks, and support of social causes – and the potential impacts of a special event on them. In doing so, it discusses ways in which event organisers and/or local governments may be able to leverage such an event in order to have intentional and positive impacts on these social assets.

This article begins with a review of relevant literature. Next, it develops a conceptual framework for studying a city's image and other social assets and then discusses how a special event can impact these social assets. Finally, it presents a case study of the Cincinnati Flying Pig Marathon, including results from two participant surveys (2002 and 2008) and one volunteer survey (2012).

2 LITERATURE REVIEW

Scholars and observers have recognised city image as a key asset that contributes to the community's ability to attract businesses and individuals (e.g., Schmid, 2005; Bradley et al., 2002). "At the dawn of the Global Century, business leaders say, the international image of a city and its region has never been more crucial for recruiting new business talent and opportunity" (Schmid, 2005). But what is city image? Lee (2006, p.367) defines it as "the subjective view or perception of a city". Morgan and Pritchard (1998) describe image as the 'currency of culture' in that it reflects and reinforces particular shared meanings, beliefs and values about the city. The concept of city image is related to the concept of brand. A brand is a name and/or symbol that has the dual purpose of identifying a good or service of a particular firm and differentiating them from those of other competing firms (Aaker, 1991). Generally speaking, a brand is designed to positively influence the image of a good or service – or the shared perceptions or views among consumers and potential consumers. Therefore, a city brand could be defined as the name and/or symbol that have the dual purpose of identifying the city and differentiating it from other cities in an effort to enhance its image.

Cities have initiated a number of different strategies in order to enhance their image directly or indirectly. Three of the traditional strategies to enhance city image are:

1 marketing, advertising and public relations campaigns

2 development and enhancement of physical assets

3 attraction and retention of professional sport teams.

First, cities have to enhance their image through marketing, advertising and public relations campaigns. For example, the city of Cancun, Mexico used marketing and advertising to try to change its image from one marked by "wet T-shirt competitions and cheap beer and tequila" for college students, according to an article in*The Economist.* Leaders of the tourism industry felt that spring-breakers 'hijacked the city's image' despite the fact that they constituted only 1% of the visitors or tourists per year (*The Economist,* 2005). In order to continue to attract newlyweds, eco-tourists and golfers, the city of Cancun launched a new marketing campaign that focuses on the sun, surf and golf greens. In Milwaukee, Wisconsin, business and civic leaders launched a five-year, $12 million campaign to create a new image and market it to people around the world. Spirit of Milwaukee Inc. – a non-profit organisation consisting of primarily business leaders – led the campaign. An article in *The Journal Sentinel* quoted Dean Amhaus, Spirit of Milwaukee president, as saying about people outside of the city, "It was not so much that they had negative perceptions of Milwaukee. They had no perceptions" (Schmid, 2005).

Secondly, cities have tried to enhance their image through the construction and enhancement of buildings and other physical assets, including convention centres, museums, and sport and concert arenas. Many postindustrial western cities are using 'high-profile' or 'flagship' projects designed "to encourage business investment from outside the locality and to upgrade the image of the city to potential investors and tourists" [Lee, (2006), p.368]. Richards and Wilson (2004, p.1931) assert that "signature buildings frequently feature in urban strategies to develop an image or 'brand' and create competitive advantage, often at great financial cost". Recent examples, they say, are the Guggenheim museum in Bilbao, Spain, and the Tate Modern gallery in London. Leaders in Newark, New Jersey, also made a big investment in physical assets with the hope that it will improve the city's image. The city paid about $210 million for the construction of the Prudential Center, and the owners of the New Jersey Devils paid the balance of $180 million. Newark Mayor Cory Booker called the arena 'the new face of Newark' (Townes, 2007). "Many residents of Newark, as well as city officials, hope the new Prudential Center will help restore the sullied image of Newark as a dangerous and crime-ravaged city with a bleak economic future.... Others believe that the new arena will give the city an economic jolt and to the residents, hope" (Townes, 2007).

Thirdly, some cities try to enhance their image is by attracting and retaining professional sport teams, particularly major-league football, baseball and basketball teams. These strategies related to sport teams can be intertwined with strategies related to physical assets, such as public financing of the construction or renovation of a facility. But there are other sport team strategies that are not linked to physical assets, such as direct cash payments to professional sport team owners. "[T]he presence of a sport team can play a role in a city's efforts to enhance its image. This can increase local pride in the community and it can be useful for attracting industry and tourists" [Sparvero and Chalip, (2007), p.13]. One empirical study found that a National Football League (NFL) team enhanced the perception of quality of life by residents of the host city (Johnson et al., 2007).

But there is some indication that cities are moving away from these more traditional strategies of image enhancement, especially those involving physical assets and professional sport teams. Richards and Wilson (2004) say that cities are disengaging to some degree from 'infrastructure-based strategies' – those that focus on architecture and physical space – because they are expensive and inflexible. In addition to direct costs, Sparvero and Chalip discuss the opportunity costs, or the forgone value or benefits of alternative strategies. "When public funds are spent to attract or retain a professional team, those funds are unavailable for education, recreation, libraries, or other projects that could benefit a larger number of residents" [Sparvero and Chalip, (2007), p.2]. Most importantly, claims of economic benefits of professional sport teams have been 'widely discredited', according to Sparvero and Chalip. For example, the Cincinnati Bengals stadium cost county taxpayers more than $350 million to build in 2000 and about $30 million per year to operate, which is 11% of the county's annual budget. "The Bengals had said that with a new stadium, the team's revenue would increase, allowing it to sign better players, win more games and attract more fans to the area. ... [T]he team has managed just two winning seasons in the new facility. Its attendance levels have actually dropped" (Albergotti and McWhirter, 2011).

Leaders in some cities are turning to special events as a less costly and more flexible way to enhance city image. Richards and Wilson (2004, p.1932) assert that events "have emerged as a means of improving the image of cities, adding life to city streets and giving citizens renewed pride in their home city". Hall (1992, p.14) says that "major events can have the effect of sharing an image of the host community or country, lending to its favourable perception as a potential travel destination". The value that some city leaders place on such events, he says, is apparent in their fierce competition to host some of them, such as the Olympic Games and the World Expo.

Studies of the impact of an event on city image have typically been done in the context of large-scale 'mega-events', such as the Olympics Games and the World Cup Final (Higham, 1999). But scholars are beginning to look at the impact of festivals and other 'non-mega' events on city image. Higham asserts that recurring small-scale events have the potential to provide the community with more net benefits than one-time mega-events or 'hallmark' events because they require fewer public resources and may be less prone to crowding and congestion. For example, Puczko and Ratz (2001) studied the Budapest Spring Festival in Hungary found that overall festival participants had a more positive image of the city of Budapest than non-festival-goers. Also, Richards and Wilson (2004) studied the Cultural Capital of Europe (CCE) event in 2001 and they concluded that Rotterdam's image was enhanced among both local residents and non-residents (i.e., visitors or tourists).

3 CONCEPTUAL FRAMEWORK: THE SOCIAL IMPACTS OF A SPECIAL EVENT ON THE HOST CITY

The conceptual framework of this article goes beyond looking at city image as the only social asset or resource, which has been the approach of most research to this point. It extends the perspective to recognise other social assets or resources of a city, using relevant work by Sparvero and Chalip (2007) as a theoretical foundation. These authors examined various impacts – economic and non-economic (i.e., social) – of professional sport teams on the cities in which they are located. In doing so, they theorise that city leaders can reap greater benefits if they *leverage* their professional sport teams rather than simply *host* them. In other words, city leaders should actively and strategically integrate a professional sport team into their economic development plans and neighbourhood redevelopment plans, not just passively wait for a team to produce economic and social benefits. Their conceptual framework identifies three categories of potential benefits that a sport team and arena can bring to the host city:

1 Economic benefits that include:

  a new skilled and unskilled jobs, possibly introducing first-timers to employment

  b tax base enhancement (income, sales and property tax)

  c redevelopment of the neighbourhoods around the venue.

2 Place marketing benefits that consist of:

  a enhancing the city image or brand by raising awareness and creating associations with desired characteristics

  b increasing local residents' pride in the city and attracting them to the venue and surrounding businesses

  c attracting businesses and workers to the city

  d increasing the number of visitors or tourists, lengthening their stay and increasing their spending.

3 Social welfare benefits that include:

a building community and strengthening social networks

b supporting of a social cause or causes.

Since this article examines the impact of a special event on the *social* assets of a city, it incorporates relevant concepts from two categories of potential benefits identified by Sparvero and Chalip: place marketing (#2 above) and social welfare (#3 above). Specifically, it uses two of the four elements in place marketing category that are non-economic – or social – in nature and all of the elements in social welfare category. The non-economic place marketing elements are:

a enhancing the city image or brand

b increasing local residents' pride in the city and attracting them to the venue and surrounding businesses.

The conceptual framework is shown in Figure 1 and discussed.

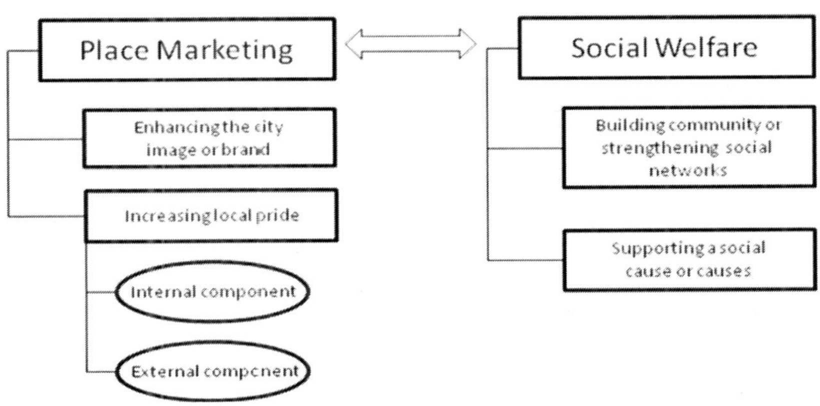

Figure 1 Potential social impacts of special events

3.1 ENHANCING CITY IMAGE

As discussed earlier, city image has been defined as shared perceptions or views of a city (Lee, 2006; Morgan and Pritchard, 1998), and city brand has been defined as the name and/or symbol that identifies a particular city and distinguishes it from other cities (Aaker, 1991). This article asserts that event organisers and/or local governments can leverage a special event – especially a recurring special event – in a way that results in the city's brand capturing desired or positive associations from the event's brand, based on Sparvero and Chalip's (2007) work. If the event attracts positive attention in newspapers and on television news programmes in the region, for example, then this may result in enhancement of the city's brand. One of the fundamental requirements of this leveraging opportunity is that the event must be associated with the host city name (Sparvero and Chalip, 2007). Examples of special events that are tied to a city are: the Boston Marathon, the New York City Marathon and the New Orleans Jazz Festival. Often times, though, the potential benefits of branding are lost when the event is not paired with a place or it is tied to a place that is very broad and diffuse. Examples of events that are not directly associated with a city include: the Rock and Roll Marathon (in San Diego, Seattle, and a number of other cities), the Sawdust Art Festival (in Laguna Beach, California), and the West Beach Music and Arts Festival (in Santa Barbara, California).

3.2 Increasing local pride

A sport team – especially if leveraged properly – can have two benefits related to local residents: enhancing their pride in the city and attracting them to the venue and surrounding businesses (Sparvero and Chalip, 2007). An extension of this idea is that a special event has the potential to foster residents' pride in their city or community – that is, their satisfaction and esteem related to their city or community. The boost in community pride following an event has been referred to as the 'halo effect' (Hall, 1992) and the 'feel-good effect' (Allen et al., 2002). Eckstein and Delaney (2002) say that community pride has two components:

1 an internal component, which refers to residents' own perception of their community

2 an external component, which refers to residents' belief of how outsiders perceive their community.

The second potential benefit of a special event is it can attract locals to participate in the event and patron businesses near the event. Getting local residents to participate in an event is important because they may be more likely than non-local residents – i.e., tourists or visitors – to be repeat participants (Sparvero and Chalip, 2007).

3.3 Strengthening social networks

A sport team – again, especially if leveraged properly – can build community and strengthen social networks through the interaction that takes place during games, at pre- and post-game parties, and through fan groups (Sparvero and Chalip, 2007). Likewise, a special event brings together participants, volunteers and others for a unified purpose. Therefore, it creates opportunities for strengthening social networks – i.e., interaction and communication – and enhancing norms of solidarity – i.e., "we're all in this together".

3.4 Supporting social causes

Another way that a sport team can impact a city or community is through involvement by the team and its members in supporting a social cause or causes (Sparvero and Chalip, 2007). Normally, the team partners with a non-profit organisation or charity, "lending its celebrity status, spokespersons, and events to the cause" [Sparvero and Chalip, (2007), p.22]. Potential benefits for the non-profit organisation are increased awareness and resources generated by the team, while the potential benefits for the team are favourable press and possibly enhancement of its image (Sparvero and Chalip, 2007). These authors suggest that the sport team and non-profit organisation work together on their public communications campaign in order to leverage the positive results for the team and the non-profit. In a similar way, a special event can help to support a social cause or causes. Some special events are organised with the intent of raising funds for a particular cause. For example, the March for Babies – formerly known as WalkAmerica – is an event that takes place in about 900 cities in recent years and has raised $2 billion for March of Dimes since 1970 (Why we walk, 2012). Other events are organised for a primary reason not based on addressing social problems – e.g., a marathon running event – but they provide opportunities for participants to raise funds for social causes or help them in some other manner. For example, more than 45,000 people ran or walked the Chicago Marathon in 2012. Of these, about 10,000 'charity runners' raised millions of dollars for more than 160 local, national and global causes, including the American Cancer Society, the American Red Cross, and the Leukemia and Lymphoma Society (Charity program, 2012).

4 A CASE STUDY: THE CINCINNATI FLYING PIG MARATHON

This article begins to apply this conceptual framework to the 'real world' by examining the impact of the Cincinnati Flying Pig Marathon on the host city's image and other social assets. It begins with some background information on the city of Cincinnati, USA, and the Cincinnati Flying Pig Marathon. Then it discusses the research design for this study including the various measures used to assess the social impacts of a special event on the host city.

4.1 Background

The city of Cincinnati is located in the southwest corner of the state of Ohio, along the northern banks of the Ohio River. Founded in 1788, Cincinnati became home to a large number of Germans who immigrated during the early 1800s and beyond. Relevant to this case study is the fact that during the mid-1800s Cincinnati became the pork-processing centre of the USA. "Because of Cincinnati's association with meatpacking, the city became known as the 'Porkopolis' of the United States" (Cincinnati, Ohio, 2009). Also, relevant to this case study is a neighbourhood called Over-the-Rhine, or OTR for short. In the early 1800s, German immigrants gave it this name because the Miami and Erie Canal resembled the Rhine River in their home country. OTR has been listed in the National Register of Historic Places and the National Trust for Historic Preservation's 'Most Endangered Historic Places in America' because it has "the nation's largest contiguous collection of Nineteenth Century Italianate Architecture" (Historic preservation, n.d.). During the mid-1900s, OTR became one of the poorest neighbourhoods in Cincinnati; however, in the 1980s and 1990s, it was experiencing some gentrification and redevelopment, including restaurants, bars, unique retail shops and other small businesses.

On another front, in the 1990s, a group of runners in Cincinnati began discussing the possibility of organising a major marathon in the city. These event organisers wanted to position the race as fun rather than competitive in order to generate interest outside the traditional running community. "During brainstorming sessions, many names were suggested that had some connection to the unique qualities of Cincinnati, such as the 'River City Marathon' or the 'Seven Hills Marathon.' ... When someone in the group mentioned 'flying pigs', everyone laughed, and the organizers knew they were on to something" [Olberding and Jisha, (2005), p.194]. The 'Flying Pig' title has historical and cultural meaning for Cincinnati because, as mentioned earlier, the city had been known as 'Porkopolis'. In addition, this title has meaning for a marathon event because people faced with a seemingly impossible goal – like running 26.2 miles – sometimes use the expression "I'll do it when pigs fly".

Flying Pig organisers created a non-profit organisation called Cincinnati Marathon Inc. to manage the marathon. The first marathon was launched in 1999 with a strong branding and marketing campaign (Olberding and Jisha, 2005). At the time, the course started in downtown Cincinnati and took runners into the historic cities of Newport and Covington in Northern Kentucky – across the Ohio River – and through some neighbourhoods in Cincinnati and suburbs; however, it did not go through OTR.

In April 2001, the city of Cincinnati faced a serious crisis. Civil unrest, or riots, occurred in the city for two days after a white Cincinnati police officer shot and killed a 19-year-old black man. The activity was concentrated in OTR, and the local and national media provided stories and images of burning and looting. The immediate damage to buildings and other property was estimated at $3.6 million. Also, a coalition of organisations called for a boycott of all businesses in downtown Cincinnati. A number of celebrities cancelled speaking engagements and concerts, including Bill Cosby, Whoopi Goldberg and Wynton Marsalis. In the first year of the boycott, the loss in convention and entertainment revenue was estimated at more than $10 million.

For the Cincinnati Flying Pig Marathon in 2002, event organisers had time to consider the course in light of the riots and the negative media coverage of the city. They decided to re-route the course to go through OTR so that participants would experience the neighbourhood first-hand and see its architecture and people as a way to counter the negative news stories, videos and photos about the riots. With hindsight, it appears that Flying Pig organisers tried to leverage the event so that it would have a positive impact on participants' perceptions of the city of Cincinnati – that is, its image – as well as positive impacts on other social assets of the city (Sparvero and Chalip, 2007).

Over the years, the Flying Pig has grown into a weekend-long event with the addition of new events including a 10-kilometer race in 2003, a 5-kilometer race in 2004 and a half-marathon in 2006. The total number of participants increased from 6,000 in the inaugural year of 1999 to more than 30,000 participants in recent years. About half of participants come from inside of the Greater Cincinnati metropolitan area – i.e., local participants – and the other half come from outside of it – non-local or visiting participants.

4.2 Research design

The purpose of this case study is to begin to assess the impacts of a special event on the host city's social assets, including city image, local pride, social networks and support of social causes. Measuring these 'soft' social impacts of a special event – primarily through participants' perceptions and opinions – is more challenging than measuring the 'hard' economic impacts, such as participants' spending on hotel rooms, food and entertainment. In order to attempt to measure social impacts, through participants' perceptions and opinions, this study utilises a unique set of data coming from two surveys of Flying Pig participants (i.e., runners and walkers) conducted in 2002 and 2008 and a third survey of volunteers conducted in 2012.

The Flying Pig organisers have administered the participant survey since the first year of the event in 1999. Originally, the survey was designed to obtain feedback from runners and walkers on their experiences to use for the planning and organising of the event in the future, and it was in paper-and-pencil format. Beginning in 2003, Burke Inc. – a marketing research company with expertise in research design, execution and analysis – began conducting the participant survey via the internet. Also, organisers have added items designed to estimate the economic impacts of the event on the city of Cincinnati.

In 2012, the authors of this article administered a survey of Flying Pig volunteers, which was the first time for a survey of this population. Some items were pulled from the 2002 and 2008 Flying Pig participant survey. The survey instrument also included a few items based on the development of the conceptual framework regarding the potential impacts of a special event on a host city's social assets. (In addition, this survey instrument included a number of items related to volunteer motivation and satisfaction. The results are being used in a different study specifically focusing on these topics.) The volunteer survey was administered online using SurveyMonkey, based on the seminal work on survey design and implementation by Dillman et al. (2008; Dillman, 2000). The Flying Pig volunteer coordinator sent a series of three messages to potential respondents, including a pre-notification, a notification and a post-notification (i.e., reminder or follow-up).

4.3 Measures of social impacts

This section discusses the various measures of social impacts in greater detail, including city image, local pride, social networks and support of social causes.

City image: The earliest attempt to measure the Flying Pig's impact on city image was in 2002, which was the year that the event organisers re-routed the event through OTR. A few simple items about city image were added to the survey of *local* participants. Specifically, the survey instrument included a question that asked local participants the following: "Did your participation in the marathon change your perceptions about downtown Cincinnati?" The survey instrument gave local respondents the opportunity to mark 'yes' or 'no' and also to write a narrative response. A total of 280 local participants responded to this survey.

In 2008, the authors worked with Burke Inc. to add two items about city image to the web-based survey not only for local participants but also for non-local participants in either the marathon (26.2 miles) or the half-marathon (13.1 miles). Each item was based on an 11-point Likert scale with 1 being 'extremely negative impact', 6 being 'no impact' and 11 being 'extremely positive impact'. (The items used 11-point Likert scales for consistency, as Burke used this format for other items on the participant survey). They were:

1 the degree to which the Flying Pig impacted your perception of downtown Cincinnati

2 the degree to which the Flying Pig impacted your likelihood to spend time in downtown Cincinnati in the future – e.g., to eat at a restaurant, attend a sporting event, visit a museum, etc.

More than 500 participants in the 2008 Flying Pig marathon or half-marathon responded to these items about city image.

Again, in 2012, the authors conducted a survey of Flying Pig volunteers. In terms of city image, the instrument included the same two items on city image that were on the 2008 participant survey – i.e., the Flying Pig's impact on their perception of Cincinnati and the likelihood they would spend time in Cincinnati. This instrument included one additional item: the degree to which the Flying Pig impacted the overall image of Cincinnati. In the end, more than 1,000 volunteers responded to this survey.

Local pride: In 2008, the authors requested that Burke researchers add two items designed to measure local pride, based Eckstein and Delaney's (2002) assertion that local pride has an internal component and an external component. The following two items were on an 11-point scale with 1 being 'extremely negative impact', 6 being 'no impact' and 11 being 'extremely positive impact':

1 the degree to which the Flying Pig impacted your sense of pride in being a resident of Cincinnati

2 the degree to which the Flying Pig impacted visiting participants' perceptions of downtown Cincinnati.

(Again, the items used 11-point Likert scales in order to be consistent with other items on the participant survey). In addition, the 2012 survey of Flying Pig volunteers included the same two items designed to measure local pride.

Social networks: The conceptual framework discusses that a special event has the potential to create opportunities to strengthen social networks and for enhancing norms of solidarity. This study begins to measure the impacts on social networks by accounting for the number of participants, volunteers and spectators that engage themselves in the Flying Pig event. In addition, the 2012 survey of Flying Pig volunteers included two items that were designed to provide other measures of the event's impact on social networks and social capital. The items asked respondents to assess the Flying Pig's impact on the following (again, using an 11-point scale):

1 social networks – interaction and communication – among individuals and groups

2 feelings of community and solidarity – i.e., "we're all in this together".

Support of social causes: Finally, the conceptual framework asserts that a special event has the potential impact of supporting a social cause or causes, which is generally represented by support of non-profit organisations. This study begins to measure these impacts using data from secondary sources on the number of volunteers and amount of charitable funds raised through the Cincinnati Flying Pig Marathon. For example, information on the number of volunteers and amount of money raised for local charities came from newsletters and other publications by the Flying Pig Marathon.

5 RESULTS

5.1 City image

In response to a question to local participants on the 2002 survey, 92 of the 280 respondents (33%) indicated that their participation in the Flying Pig event changed their perceptions of downtown Cincinnati. And 79 of the 92 respondents (85.9%) said it had a positive impact, while only two of them (2.2%) said it had a negative impact. The results are shown in Table 1.

Local participants who responded to the 2002 survey were provided the opportunity to write narrative responses explaining their perceptions of downtown Cincinnati. Table 2 provides some sample comments by attribute. For example, one respondent wrote: "The marathon is such a friendly, approachable event that takes place in our city center that it overrides many of the bad impressions I had had recently about Cincinnati". And another one wrote: "A little more pride in the city, and a feeling we are 'big league' in more than just pro sports".

Table 1 Impact of the 2002 Flying Pig on local participants' perceptions of downtown Cincinnati

	Local participants	
	Number	*Percentage*
Positive	79	85.9%
Negative	2	2.2%
Both positive and negative	1	1.1%
Neutral	10	10.9%
Total	92	100.0%

Table 2 Sample qualitative statements by local residents who participated in the 2002 Flying Pig

Lots of places to see and things to do – restaurants, shopping, etc.

I have never really had the chance to go downtown and with everything you hear on the news, I wasn't in a hurry to go. Since I had a chance to see downtown better because of the marathon, I will probably go there in the near future. I didn't realise there were so many unique shops.

I never utilised the downtown restaurants until I ran by them and found where they were. I very much like to go downtown now due to the marathon.

I am new to the area and I never knew what the downtown really had to offer. Since the Marathon we have gone to Music Hall, twice and to a restaurant.

Friendly and hospitable people – community support

The way people came together was great – and it made me realise that I could get that feeling by attending other events downtown.

Local neighbourhood people that ran the checkpoints seemed very friendly

Safe/comfortable

My friends and I stayed at a downtown hotel the night before the marathon and went out for dinner and a drink. There were several high school proms being held downtown and it was fun to people watch. Overall, downtown seemed more safe than I have thought it to be in the past.

Physical beauty

I realised how beautiful Cincinnati really is and how it is taken for granted. I visited several neighbour[hoods] that I would never have gone to if I had not ran the Flying Pig Marathon. I always thought some areas were dangerous, but I found out otherwise. ... The areas are quite nice and they are in the process of being renovated and improved.

It gave me a chance to see the improvements that had been made to the downtown area and the riverfront area.

Easy to move around and park

Ease to get around, friendliness of people.

It's a lot nicer and more accessible than I had thought.

Clean

Some areas were much cleaner than I expected.

Diverse

I am a Cincinnati resident and felt more comfortable being downtown than in the previous few months. It was nice to see a diverse crowd united for the event.

General positive comments

I was pleased that out-of-town participants thought that Cincinnati was – as I believe – a much better city than they had been led to believe by national press coverage after the April 2001 riots.

A little more pride in the city, and a feeling we are 'big league' in more than just pro sports.

For the 2008 survey, the city image items were given not only to local participants but also to *non-local or visiting* participants. Table 3 displays the mean responses to general questions about the impacts of the Flying Pig event on city image including:

1 participants' perception of downtown Cincinnati

2 the likelihood that they will spend time in downtown Cincinnati in the future.

A key finding is that the Flying Pig Marathon Weekend had a very positive impact on local residents' perception of downtown Cincinnati (mean of 8.53 on an 11-point scale). This event had an even more positive impact on non-local participants (mean of 9.33). Further, the results indicate that the Flying Pig had a somewhat positive impact on the likelihood that local participants would spend time in downtown Cincinnati in the future (mean of 7.66) and an even stronger impact on non-local participants (mean of 8.76).

Table 3 Participants' assessment of the Flying Pig's impact on Cincinnati image (2008)

	Means (standard deviations)*	
	Local participants	*Non-local participants*
The Flying Pig's impact on their own perceptions of downtown Cincinnati	8.53 (1.83)	9.33 (1.63)
Number of respondents (n)	229	298
The Flying Pig's impact on the likelihood of their spending time in downtown Cincinnati in the future	7.66 (1.77)	8.76 (1.79)
Number of respondents (n)	229	297

Notes: *11-point Likert scale with 1 being 'extremely negative impact', 6 being 'no impact'.

Table 4 Volunteers' assessment of the Flying Pig's impact on Cincinnati image (2012)

	Means (standard deviations)*	
	Local volunteers	*Non-local volunteers*
The Flying Pig's impact on their own perceptions of Cincinnati	9.02 (1.46)	9.04 (1.40)
Number of respondents (n)	937	53
The Flying Pig's impact on the likelihood of their spending time in Cincinnati in the future	8.47 (1.99)	8.94 (1.81)
Number of respondents (n)	955	53
The Flying Pig's impact on the overall image of Cincinnati	9.09 (1.55)	8.69 (1.73)
Number of respondents (n)	947	52

Notes: *11-point Likert scale with 1 being 'extremely negative impact', 6 being 'no impact' and 11 being 'extremely positive impact'.

Recall that similar items were included on the 2012 survey of volunteers and, interestingly, results are quite similar those from the 2008 survey of participants. As shown in Table 4, volunteers indicated that the Flying Pig Marathon Weekend had a very positive impact on their perception of Cincinnati (mean of 9.02 on an 11-point scale for local volunteers and 9.04 for non-local volunteers). The results also indicate that the Flying Pig had a positive impact on the likelihood that volunteers would spend time in Cincinnati in the future (mean of 8.47 for local volunteers and 8.94 for non-local volunteers).

The volunteer survey had one additional item related to city image, which asked respondents about the Flying Pig's impact on the overall image of Cincinnati. Similar to the participants' responses, volunteers indicated that this special event had a positive impact on city image (mean of 9.09 for local volunteers and 8.69 for non-local volunteers).

5.2 Local pride

In addition to items about local and non-local participants' perceptions of the city, the 2008 survey had two items for local participants regarding city pride. One item was designed to tap into the internal component of local pride (Eckstein and Delaney, 2002), which refers to residents' own perceptions of their community. As the first column of Table 5 indicates, the mean response was 9.17 (on an 11-point scale), indicating that the Flying Pig Marathon had a very positive impact on local participants' pride about Cincinnati. A second item was designed to tap into the external component of local pride (Eckstein and Delaney, 2002), which refers to residents' belief of how outsiders perceive their community. The mean response was 7.66, indicating that local residents believed that the Flying Pig Marathon had a somewhat positive impact on visitors' perceptions of Cincinnati.

Table 5 Impact of the Flying Pig Marathon on local pride

	Means	
	Local participants (2008)	*Local volunteers (2012)*
Assessment of the Flying Pig's impact on their own sense of pride about living in Greater Cincinnati (internal component)	9.17 (1.64)	8.98 (1.55)
Number of respondents (n)	229	952
Assessment of the Flying Pig's impact on visitors' perceptions Cincinnati (external component)	7.66 (1.77)	9.02 (1.41)
Number of respondents (n)	228	940

Notes: *11-point Likert scale with 1 being 'extremely negative impact', 6 being 'no impact' and 11 being 'extremely positive imapct'.

Again, similar items were included on the 2012 survey of volunteers and, again, results echo those from the 2008 survey of participants. As the second column Table 5 indicates, Flying Pig volunteers indicated that this event had a positive impact on their own sense of pride of living in Greater Cincinnati – i.e., the internal component – and on their beliefs about visitors' perceptions of the region – i.e., the external component. These means are 8.98 and 9.02, respectively.

5.3 Social networks

The Cincinnati Flying Pig Marathon engages a number of people throughout the weekend. In recent years, more than 30,000 people participated in the running/walking events: the marathon (as individuals or as members of four-person relay teams), the half-marathon, the 10-kilometer race, the 5-kilometer race and the kids' fun run. It should be noted that individuals train together for months prior to the event – usually at least once a week – either informally (i.e., with friends, neighbours, co-workers, etc.) or formally (i.e., through a training group or programme set up by a local running shop or non-profit organisation). Also, the relay teams tend to be sponsored by a corporation for which the four members work or by some other organisation for which they are members. The training prior to the race and running together on the day of the event tend to build community – i.e., enhanced interaction and communication – and a sense of solidarity – i.e., "we're all in this together". This community building and social networking also occurs at other non-running events during the weekend, such as the pasta party, which takes place the night before the marathon and half-marathon.

In addition to the runners and walkers, this event involves about 4,000 volunteers who fill and distribute runners' bags, pass out water along the course and provide first-aid when needed (The squeal, 2009). Further, spectators line up along the 26.2-mile course spectators to cheer on the runners and the walkers, which creates additional community-building and social-networking opportunities, similarly to fans at a sport event (Sparvero and Chalip, 2007). In recent years, there have been an estimated 150,000 spectators along the course (The squeal, 2009).

In the 2012 survey of volunteers, an effort was made to quantitatively measure the Flying Pig's impact on social networks. Respondents indicated that the Flying Pig had a positive impact on social networks (mean of 8.40 for local volunteers and 8.21 for non-local volunteers). In addition, they indicated this event had a positive impact on feelings of community and solidarity (mean of 8.68 for local volunteers and 8.45 for non-local volunteers).

5.4 Support of social causes

More than 100 non-profit organisations benefit from the Cincinnati Flying Pig Marathon, including national organisations like the

Leukaemia and Lymphoma Society and the American Cancer Association as well as local organisations such as the Cincinnati Museum Center, Hospice of the Bluegrass, a few branches of the YMCA, various Boy Scout troops, and a number of elementary and high schools (Charity listing, n.d.). Some of these organisations provide volunteers to staff areas; in exchange, Cincinnati Marathon Inc. makes financial contributions to these profits. In addition, a number of non-profit organisations use the Flying Pig Marathon as an opportunity to raise funds for their organisations; 'charity runners' and 'charity walkers' select a non-profit and then get family and friends to sponsor them and donate funds to their cause. About $9 million has been raised for charitable organisations through the Flying Pig Marathon since its inception in 1999 (Charity listing, n.d.).

Table 6 Impact of the Flying Pig Marathon on social networks (2012)

	Means (standard deviations)*	
	Local volunteers	*Non-local volunteers*
Assessment of the Flying Pig's impact on social networks – interaction and communication – among individuals and groups	8.40 (1.68)	8.21 (1.84)
Number of respondents (n)	947	53
Assessment of the Flying Pig's impact on feelings of community and solidarity – i.e., "we're all in this together	8.68 (1.58)	8.45 (1.79)
Number of respondents (n)	951	53

6 CONCLUSIONS

This article builds upon emergent concepts and theories related to the relationship between a special event and the host city's social assets. In doing so, the conceptual framework takes a broader perspective of social assets by including not only city image but also local pride, social networks, and support for social causes. The conceptual framework is based on one that was developed to study the economic and social impacts of professional sport teams on host cities (Sparvero and Chalip, 2007). This framework is adapted to study the potential social impacts of special events on host cities. Further, the city image literature is used to build upon this framework by expanding the definition of city image to include external and internal components (Eckstein and Delaney, 2002). Ultimately, these theories and concepts are used to develop new measures of these social assets, which are then applied in a case study of the Cincinnati Flying Pig Marathon.

Of course, it should be noted that there are limitations to this study. First and foremost, the empirical portion of this article is a case study focusing on one special event in one host city in one country. Thus, a next logical step in this research programme would be to conduct similar studies of other events in other cities in the USA and beyond. Another limitation of this study is that it does not have a control group. The research design would be more robust with both an experimental group consisting of individuals who participated in a special event and a control group consisting of individuals who did not. Again, this may be another direction for future research, although it may be challenging to implement. In any case, additional studies of the social impacts of special events would be most informative and insightful if a similar conceptual framework and measures are used. That way, results could be more readily compared, contrasted and cumulated to better determine the degree to which findings are consistent and whether there are patterns or trends in them.

Moving from a discussion of the academic contributions and limitations, this study may help event organisers and city leaders in the 'real world' to better understand the ways that a special event can impact a community, especially its social assets. The case study finds evidence that the Cincinnati Flying Pig Marathon has enhanced city image among local and non-local participants; increased local participants' own sense of pride about the city and raised their perceptions of how visitors perceive Cincinnati; created opportunities for social interaction among more than 30,000 runners and walkers, 4,000 volunteers and 150,000 spectators; and raised nearly $9 million for social causes. The positive impacts of this special event on the host city's social assets are noteworthy in light of the low amount of public funds necessary to support it, especially relative to more traditional strategies to improve city image and other social assets, such as capital projects and marketing campaigns. Specifically, the cost to the city of Cincinnati to support the Flying Pig Marathon is about $100,000 per year, primarily for police and EMS services. In contrast, as mentioned earlier, the cost to county taxpayers to support the Cincinnati Bengals was more than $350 million in 2000 for stadium construction and about $30 million per year for stadium operations (Albergotti and McWhirter, 2011). The story of the Cincinnati Flying Pig Marathon may offer a lesson or, at least, a point of consideration for event organisers and city leaders in Cincinnati and beyond: The cost of – or investment in – a special event can be relatively low while the positive impacts on both social and economic assets of a city – or return on investment – can be relatively big. A related lesson from the literature (Sparvero and Chalip, 2007) is that event organisers and city leaders should approach special events strategically and collaboratively in order to leverage it and obtain the greatest possible positive impacts, or return on investment.

The authors would like to thank Burke Inc. for including items about social impacts to surveys of Flying Pig participants as well as Elizabeth Engelhardt for helping to develop the survey of Flying Pig volunteers.

AUTHOR DETAILS

Julie Cencula Olberding is the Director of and Associate Professor of the Master of Public Administration (MPA) programme at Northern Kentucky University. Her research has focused on non-profit organisations and management, philanthropy and volunteerism, and economic and community development. It has been published in *Public Administration Review, Economic Development Quarterly, International Journal of Public Administration, International Studies Perspective, Journal of Public Affairs Education* and *Innovative Higher Education*. She obtained her MPA and PhD from the Martin School of Public Policy and Administration at the University of Kentucky and her Bachelor's degree from the EW Scripps School of Journalism at Ohio University.

Douglas J. Olberding is the Chair of and Associate Professor at the Department of Sport Studies, Xavier University. His research has focused on the economic impact of sport events, branding of sport events, and strategic planning and management of sport organisations. His articles have been published in the*International Journal of Sport Management, International Journal of Sport Finance, Sport Marketing Quarterly,* and *International Studies Perspective*. He obtained his doctoral degree in Sport Management from the University of Kentucky, a Master's degree in Political Science from Ohio University, and a Bachelor's degree in Political Science from Xavier University.

REFERENCES

1 Aaker, D.A. (1991). Managing Brand Equity. New York:Free Press

2 Albergotti, R. , McWhirter, C. 'A stadium's costly legacy throws taxpayers for a loss'. The Wall Street Journal.2011, 07, 12, (accessed 6 August 2012), [online] http://online.wsj.com/

article/SB10001424052748704461304576216330349497852.html
3 Allen, J. , O'Toole, W. , McDonnell, I. , Harris, R. (2002). Festival and Special Event Management.London:Wiley
4 Belson, K. 'Marathon organizers highlight city's economic gain'. The New York Times. 2011, 04, 27,(accessed 6 August 2012), [online] http://www.nytimes.com/2011/04/ 28/sports/28 marathon.html?_r=0
5 Bradley, A. , Hall, T. , Harrison, M. (2002). 'Selling cities: promoting new images for meetings tourism'. Cities.19, 1, 61-70
6 Charity listing (accessed 17 November 2011), (n.d.) [online] http://www.flyingpigmarathon.com/sponsors_and_charities/charity_listing.shtml
7 Charity program (2012). (accessed 17 November 2011), [online] http://www.chicagomarathon.com/charity-community/charity-program/
8 Cincinnati, Ohio (2009). (accessed 1 March 2009), [online] http://www.ohiohistorycentral.org/entry.php?rec=681
9 Cincinnati: Population Profile (2009). (accessed 1 March 2009), [online] http://www.city-data.com/us-cities/The-Midwest/ Cincinnati- Population-Profile.html
10 Cobb, S. , Olberding, D.J. (2007). 'The importance of import substitution in marathon economic impact analysis'. International Journal of Sport Finance. 2, 2, 108-119
11 Cobb, S. , Olberding, D.J. (2008). Economic Impact Analysis of the Flying Pig Marathon. Cincinnati, Ohio
12 Cobb, S. , Olberding, D.J. (2012). Economic Impact Analysis of the 2012 Flying Pig Marathon. Cincinnati, Ohio
13 Crompton, J. (1995). 'Economic impact analysis of sports facilities and events: eleven sources of misapplication'. Journal of Sport Management. 9, 1, 14-35
14 Dillman, D.A. (2000). Mail and Internet Surveys: The Tailored Design Method. 2nd ed., New York:John Wiley and Sons

15 Dillman, D.A. , Smyth, J.D. , Christian, L.M. (2008). Internet, Mail and Mixed-Mode Surveys: The Tailored Design Method. 3rd ed., New York:John Wiley and Sons

16 Eckstein, R. , Delaney, K. (2002). 'New sport stadiums, community self-esteem and community collective conscience'. Journal of Sport and Social Issues. 26, 3, 235-247

17 Hall, C.M. (1992). Hallmark Tourist Events: Impacts, Management, and Planning. London:Belhaven Press

18 Higham, J. (1999). 'Commentary – sport as an avenue of tourism development: an analysis of the positive and negative impacts of sport tourism'. Current Issues in Tourism. 2, 1, 82-90

19 Historic preservation (accessed 1 March 2009), (n.d.) [online] http://www.otrfoundation.org/historic_preservation.php

20 Johnson, B.K. , Mondello, M.J. , Whitehead, J.C. (2007). 'The value of public goods generated by a National Football League team'. Journal of Sport Management. 21, 1, 123-136

21 Johnston, K. 'Boston Marathon to generate $137.5 million'. The Boston Globe. 2012, 04, 12, (accessed 6 August 2012), [online] http://www.baa.org/news-and-press/news-listing/2012/ april/ boston-marathon-weekend-economic-impact.aspx

22 Lee, C.B. (2006). 'High profile projects and tourism policy in Birmingham, England: do they work to enhance the city's image and promote inward business investment?'. Planning, Practice & Research. 21, 3, 367-381

23 Morgan, N. , Pritchard, A. (1998). Tourism, Promotion and Power: Creating Images, Creating Identities.Chichester:John Wiley and Sons

24 Olberding, D.J. , Jisha, J. (2005). "The Flying Pig': building brand equity in a major urban marathon'. Sport Marketing Quarterly. 14, 3, 191-196

25 Porter, P. , Fizel, J. , Gustafson, E. , Hadley, L. (1999). 'Mega-events as municipal investments: a critique of impact analysis'. Sports Economics: Current Research. Westport:Praeger , 61-73

26 Puczko, L. , Ratz, T. , Richards, G. (2001). 'The Budapest Spring Festival: a festival for Hungarians?'. Cultural Attractions and European Tourism. CABI
27 Richards, G. , Wilson, J. (2004). 'The impact of cultural events on city image: Rotterdam, Cultural Capital of Europe 2001'. Urban Studies. 41, 10, 1931-1951
28 Schmid, J. 'Paint it Milwaukee: city in transition seeks satisfaction with more contemporary image'. Journal Sentinel. 2005, 10, 01
29 Sparvero, E. , Chalip, L. (2007). 'Professional teams as leverageable assets: strategic creation of community value'. Sport Management Review. 10, 1, 1-30
30 'A break from spring break: a brave attempt at rebranding a rowdy resort'. The Economist. 2005 03 12
31 The squeal (2009). The Latest and Greatest News from the Cincinnati Flying Pig Marathon. (accessed 17 November 2011), [online] http://www.flyingpigmarathon.com/news_ flash/the_squeal/050109.shtml
32 Townes, G. (2007). 'New Prudential arena opens in Newark: will it help restore the city's image?'. New York Amsterdam News. 98, 4,
33 Why we walk (2012). (accessed 1 August 2012), [online] http://www.marchforbabies.org/why.aspx

CHAPTER 3

Terrorism, homeland safety and event management

Geoffrey R. Skoll[1], Maximiliano E. Korstanje[2]

[1]Criminal Justice Department, Buffalo State College,

1300 Elmwood Avenue, Buffalo, NY 14222, NY, USA

[2] Department of Economics, University of Palermo,

Larrea 1076 3 Floor Buenos Aires, Argentina

ABSTRACT

As the last attacks on Boston showed terrorism is based not only on speculation but also on surprise. Terrorists do not want to destroy or to kill everybody, their goal is aimed to inflict and administrate fear to the witnesses. The fact is that tourism and mega-events represented a fertile source to perpetrate terrorist attacks, not only for the casualties but also by the psychological effects on citizenry.

This paper explores the nature of terrorism in the context of leisure as well as proposing a valid model to understand the connection among tourism, event management and terrorism.

KEYWORDS

terrorism, fear, event management, tourism, leisure, rules, surprise

1 INTRODUCTION

The conceptual limit for defining tourists' risks as those aspects which may jeopardise the well-being of tourists seems to be unquestionable. As Tarlow (2011) put it, their vulnerability and their lack of familiarity make tourists a prey for crime and terrorist attacks. The prosperity of economies which adopt tourism as a chief industry depends on the image of destinations. Of almost all risks, terrorism today is the major topic that concerns policy makers in tourism and hospitality. However, it becomes a buzz word with many significations. Though terrorism represents serious problems for the industry of tourism because of its dependence on local economies, it is very hard to define this issue with accuracy. Under some conditions, peripheral countries developed a paradoxical situation. At one time, they adopt events and festivals to mitigate the problems of poverty, and by doing so, local resentment is triggered by the material asymmetries found in these events, which then present a clear target to attack. First, we have to define what we understand by terrorism and mediated events to give a fresh conceptual model to evaluate the problem.

It is interesting to discuss to what extent mega events are a fertile source for terrorism. Terrorism introduces a message in society aimed at undermining states or creating political instability. From immemorial times, human beings have managed festivals and mega-events not only to celebrate a good harvest, but also to strengthen the social bond. Mediated events are, at least, significant celebrations that contribute socially to the reproduction of society. In view of this, some groups which are antagonistic to the state, attack against attendees at the events as a way to create panic, undermining not only the legitimacy of government but the trust of citizens in their institutions. This double effect has been widely studied by sociologists and anthropologists in terrorism related issues. However, less attention has been given to the fact that leisure spaces are preferred by terrorists to plan their acts. This raises an interesting point of debate based on the connection between terrorism, leisure, and mega-events that deserves to be continued, what is terrorism? What terrorists opt to target festivals and events? Is the nation state able to prevent the next attack? Why does leisure represent a fertile source for terrorism? The present essay explores the literature on terrorism and hospitality, and focuses on the importance leisure spaces have for terrorists. Our goals are aimed at discussing the diverse definition of terrorism as well as proposing a theory to understand its effects on social imaginary. Terrorism works by the psychological fear it can instil on targeted populations.

Effects of terrorism in public spectacles depend on four variables.

- Threshold of control: this factor exhibits the degree of control the society exerts on the event. For example, bombing can be more easily and visually monitored than food contamination by micro-organisms.

- Probability of repetition signals that the troubling event can be repeated in a short run. This means that the state lacks the necessary resources to prevent a similar attack in the future.

- Targeted victims refer to the status of casualties. Societies valorise some groups over others. Whenever targeted victims represent children or vulnerable women, because they are the biological background of community, people panic.

2 PRELIMINARY DEBATE

Although terrorism has been posed as a serious challenge for the West in last few decades, the fact is that the misunderstanding of what it means is rising. In view of this, Moten (2010) argues that terrorism is colourised by ideology or location. As an instrument of struggle, it connotes a group of rebels and a state which attempts to impose its hegemony. Summarising the four waves of Rappoport, Moten said that terrorist attacks are more violent and virulent than decades ago. Although the first wave surfaced with the anarchism and its anarcho-syndicalism programmes (1878–1910), a second wave flourished in colonised Africa and Asia from 1920s onwards. Based on a specific occupied soil, rebels struggled by their independence, their attacks were directed to targeted government. Third wave, rather, was associated with the political left in its response to the decline of democracy in Western countries (1970–1980). The last wave was symbolised by the Ayatollah's revolution in Iran 1979, which focused on the radical rejection of secular values. These types of terrorism included bombing cars and suicide bombings. One of the most troubling aspects in defining terrorism is that it serves political interests. Western states call terrorists all those individuals or groups that can destabilise the societal order, while Muslims call terrorists those imperial powers which historically have taken their resources.

We offer the following as a clarification of definitional confusion. More than 30 years ago, the US Task Force on Disorders and Terrorism proposed a serviceable definition as follows. "Terrorism is a tactic or technique by means of which a violent act or threat thereof is used for the purpose of creating overwhelming fear for coercive purposes" [Task Force on Disorders and Terrorism, (1976), p.3]. Shortening it does no harm so a more concise version is "Terrorism (n.): A coercive tactic using fear through violence or its threat." The background for the national Task Force was social turmoil accompanied by an increase in civil disorders during the early 1970s. The Task Force noted numerous airplane hijackings, bombings, and riots in cities in the USA.

Just before the 9/11 attacks, the former staff director of the Task Force weighed in on the definitional controversy. Cooper (2001, p.881) begins by quoting Cohen (1990, pp.41–42). "A living language has no existence independent of culture. It is not the loom of culture but its data bank. As such, it serves the needs, past and present, of a given community. As those needs change, language evolves to accommodate them". Cooper goes on to offer an elegant definition: "... [T]errorism is the intentional generation of massive fear by human beings for the purpose of securing or maintaining control over other human beings" [Cooper, (2001), p.883]. Cooper's 2001 definition does not require violence, but stresses control more than mere coercion. These changes improve the 1976 version as they open the possibility that terrorism is a condition, not merely an event.

A short hand version allows researchers to recognise terrorism when they see it. According to that criterion, terrorism occurs when someone makes people fearful in order to control them. The three elements are intent, fear, and control (Crenshaw, 1983, 1995).

Terrorism related research has advanced in the social sciences according to problems raised in the Cold War. That was up to 9/11, where after, many studies in tourism and hospitality turned their attention to terrorist issues (Hall, 2002; Kozak et al., 2007; Kuto and Groves, 2004; Henderson, 2008; Floyd and Pennington-Gray, 2004; Korstanje, 2009; Abdel-Azim, 2010). The hospitality and tourism industry faced serious problems because of the effects of terrorism worldwide (Baker and Coulter, 2007; Atkins et al., 2003). The symbiosis between tourism and terrorism points to the profits of the industry. Many peripheral countries adopt tourism to relieve their economic problems. Mega events provide a fresh source of income to mitigate poverty, but the problem is that these policies should be accompanied with other developmental projects to gain sustainable economic growth. The revitalisation of economies based exclusively on tourism resulted in serious imbalances for tourism-adopting countries. The modern economy of tourism and hospitality in the world of service seems to be subject to intangible experiences. Since the tourist-experience cannot be appreciated once consumed, experts rely on the question of image. The discipline of psychology has played a pivotal role in understanding tourist experiences. Given the context of subjective experience, it is not surprising to see how terrorist try to cause fear by attacking tourists. They aim at creating a message to other tourists in the world, so that they desist from travel to a particular destination. That way, states which rely on tourism, face a downfall in international travel creating a serious economic loss. Terrorism uses mega festivals or events as the platform to portray a message to the world. This situation leads policy makers and police to a paradox: the more attractive a destination is, the more dangerous it may be. The needs of planning for safety not only affect the image of destinations such as Israel or Jordan, but also create a vicious circle in case of attacks. One of the aspects that shocked the public opinion after 9/11 was not the tactic itself, which was based on western marketing and managerial manuals, but the fact it had been perpetrated in the supposed safest nation of the world. If this happened with the USA, a superpower, what could have other countries do where homeland safety appears not to be a priority of state?

9/11 undoubtedly marked a turning-point that changed the way of perceiving the world forever. The psychological impacts of 9/11 transcended the boundaries of nationhood because of the degree of visual technology developed by the media in West. Terrorists not only employed the West's own means of transport, which created a new tactic of terrorism, but also found in the media a compliant distributor for fear mongering message. Clearly, tourism has been one of the industries most affected by terrorist acts. Terrorism determines the way travellers garner information and imagine their destinations (Peattie et al., 2005). Because of their unfamiliarity with the destination, travellers and tourists are often targets of diverse crimes. Some terrorist cells attack tourists with a double message. On one hand, they inspire a panic in public opinion from the victims' countries of origin. On the other, they undermine the citizenry's trust in the state. Of course, any destination combines risk aversion with risk attraction factors. As Lepp and Gibson (2008) put it, this industry seems to be circumscribed by two contrasting tendencies: the sensation or novelty seeking risk and risk aversion. The type of psychological personality of tourists plays a crucial role at time of determining the perception of risk. In addition, West (2008) considers the terrorist attacks in 2003 to Western tourists in Bali. They have been memorialised by the Australian Press as the archetype of terrorism, comparing this event with 9/11. This means that collective memory and crises are inextricably intertwined in the national discourse. Postmodern nationalisms legitimise travel as a universal benefit to mankind which should be defended at any cost. Similarly, the narrative of terrorism emphasises that enemies of democracy utilise foreign tourists precisely because of their vulnerability, as acts of cowardice.

3 TERRORISM AND CULTURAL VALUES

As something else than a criminal act, terrorism should be understood as a political act. However, moral philosophy recently asked to what extent victims and culprits are two sides of the same coin. What for some is considered an act of cowardice for others is lived as the fight for freedom. Quite aside from this, what the 9/11 attack signified was not only the failures of control in the airspace of the USA but also the onset of a new state of fear that invaded other countries worldwide. After all, the war on terror initiated by G.W. Bush's administration engendered more problems than solutions. In this context, Kucukaltan (2006) introduces readers to the slippery connection between tourism and terrorism. His thesis represents a serious academic effort to expand the current understanding of how tourism works worldwide and the impacts of terrorism on local economies. Travel and technological advances have certainly transformed the modern forms of displacement. If one pays attention to the fact that their high degree of mobility makes international tourists an easy target of terrorist cells, it is clear how after 9/11 the relationship between tourism and terrorism became more troublesome. After further examination of the economic factors that potentiate or mitigate a state of emergency, this insightful work suggests that tourism and hospitality are sensitive to economic inequalities. Ranging from inflation and increasing of prices to the ecological deterioration of non-renewable resources, economic crisis causes several profound problems to tourism simply because it makes pathways for political instability and terrorism. The consequences of terrorism might extend even to neighbouring countries, which is of paramount importance during policy making and planning. The primary point of discussion is associated with the belief that all wars as well as political crises correspond with the convergence among religious, economic, and ideological reasons. To some extent, Kucukaltan acknowledges that even if war shows negative aftermaths for local economies, the nostalgia for involved countries or the sentiment of nationalism can very well create a demand and boom for tourism and hospitality industries.

Similarly, Bianchi (2007) explains that tourism revolves around risk perception, which acts as conducive to the interests of some industrialised nations and to the detriment of the periphery. The ongoing state of insecurity created by terrorism references a political logic of exclusion. The paradox lies in that studies of risk perception themselves threaten the goal of security they encourage. To what extent does terrorism affect the tourism industry? Castaño (2005) propones the arrival statistics from 2000 to 2003 in some cities that had been targets of terrorist attacks. This dataset questions the belief that terrorism threatens tourism. Cities like Mombasa, New York, Madrid, London, Bali, and Cairo experienced notable declines in tourism, but they recovered within a few months. Terrorism may potentiate tourism by means of dark tourism – terrorism tourism. Castaño (2005) argues that terrorism is a reversible process. No matter the original impact on public opinion, given some unspecified time frame, what today generates fear, tomorrow will entice thousands of tourists. Hotel chains and tourism staff become targets of attacks because they symbolise the strength of an economic order that causes resentment and exclusion. If the West is named as the cause of all suffering, this diminishes the responsibilities of local elites who give their support to imperialist powers.

Steiner states that marketing and management literature is ignoring the complex relationship among risk, attractiveness, and tourism. Sometimes terrorism boosts tourist destinations. What changes in these types of events is the dynamic of tourism and its fluxes, not tourism as an activity. After 9/11 many Muslim destinations declined in visitations while others prospered. To what extent may tourism be considered a retractile activity? Steiner argues that terrorism affected the Western demand in Middle East but encouraged the intra-regional markets opening new opportunities for Muslim destinations. This engendered a two sided aftermath which alternates stabilisation with conflict. In spite of negative images portrayed by the media about the 9/11 attacks, people still move toward other regions.

"the Events following 9/11 caused a recurring neighbourhood effect in various Muslim countries, which led to a decline in tourist arrivals. However, while some major destinations in the region severely suffered, others performed extremely well. From a theoretical point of view, this difference in the occurrence of a neighbourhood effect is somehow surprising and begged the question of how a varying emergence of neighbourhood effects and the varying development paths of different destination in the Muslim world after 9/11could be explained" Steiner (2010, p.203).

To respond to why terrorists select some targets while avoiding others, Vukonic (2010) explores the importance of tourism in the Middle East. The Muslim World has suffered many changes caused by the advance of modernity and tourism. Although some countries keep a friendly relationship of cooperation and diplomacy with the West, others, like Iran, adopt an atmosphere of rivalry and conflict. What remains clear for Vukonic, is that Western travellers are sometimes attacked to take advantage of their lack of familiarity with the environs they visit and because of the cultural values they represent. Unless the state intervenes in their protection, tourism may offend local tradition and customs. In this vein, Vukonic explains that the religion/tourism connection may be detached in three primary ways:

a religion somehow supports tourism

b tourism somehow influences religion

c religion and tourism stand in opposition.

It is widely accepted that religion gives support to tourism only when religious norms are not defied or in danger. The four great religions depended and depend on pilgrimages and holy land explorations. To adhere to religious doctrine, psychical displacement and travels play a crucial role because they pave the way for the construction of infrastructure, shrines, and designated holy places. Under some circumstances, tourism exerts considerable influence on religion depending on the viewpoints of residents and participants.

Some countries at odds may reinforce mutual understanding by means of tourism, even borrowing religious values from each other. Religion and tourism can be irreconcilable enemies in a way that reduces the peace promoting processes which prioritise intercultural differences rather than commonalities. Some practices introduced by guests may be incompatible with the religious beliefs of their hosts. In this case, two or more values may cause nonnegotiable discrepancies that lead groups to conflicts, and even warfare. Since tourism accelerates the intergroup connection, in such a context, conflict surfaces.

Throughout the Middle East and beyond, developing countries adopted tourism as a primary industry to adjust some economic asymmetries resulting from problems of governability, corruption, and bad administrations. Although in the short run, tourism brings fresh investments to host economies, states can develop a dependency on the tourism industry. Terrorists and insurgents may take advantage of this situation to affect the destination image. This way, they not only attack the state but also jeopardise its economy. Pro-tourist policies sometimes contrast with some radicalised beliefs. Therefore, terrorism is for Vukonic a glitch in the process of communication, an anomaly that makes tourism and religion not understand each other. Vukonic's argument is that in an intercultural meeting some groups are encouraged to improve their situations, while others do not. What is important to discuss is why terrorist attacks have been imported to the greater Western cities in recent years? Undoubtedly, 9/11 marked the starting point of a new frame of reference where citizens do not feel safe anywhere. The marathon bombing in Boston 15 April 2013 illustrates the point that public celebrations seem to be a playground for the operations of terrorists in USA cities. The problem with terrorism is that nobody knows when the next blow will take place. Terrorists today are educated in the best universities of West, some of them live legally in European or USA cities, and some are even citizens of the USA or European Union countries. The problem of terrorism differed substantially in earlier decades. To understand why terrorists select the public mega events for their attacks we first have to explore the world of mediated events and their psychological effects on society.

4 THE MEDIA EVENTS AND SAFETY

From an anthropological perspective, festivals have an epicentre in the allegoric construction of meaning in relation to certain themes which concern the involved community. Events and festivals must be understood as rites whose essences resolve existential cultural contradictions. For example, societies have serious problems in intellectualising the death of children. Because of biological cycles, they are supposed to live more than parental generations. Event management corresponds with a rite of passage that describes the economic cycle of harvest.

Festivals represent not only certain denial of poverty and shortage expressed in banquets, but also emulates lost paradise through beauty and harmony. The influence exerted by mediating events is unquestionable. In a longitudinal investigation by Ritchie et al. (2007) in the capital city of Camberra in Australia showed that viewers subjected to mediated events had a more positive image of the destination than other groups.

Liang et al. (2008) assert that event management entails three benefits associated with enjoyment, socialisation, and appreciation. Supporting the findings of previous studies, the authors conclude that festivals and events not only contribute to the expansion of local economy but revitalise the commitment for being part of these events irrespective of the travelled distance. Following this, Crompton and Mckay (1997) suggest that six motives dominate visitors at the time of participating in a festival: cultural exploration, novelty, recovery needs, socialisation, external interaction, and gregariousness. Other effects of festival management are strictly related to the image change for a community (Prentice and Andersen, 2003), politically staged authenticity issues or guest and host's conflicts (Berlanga-Adell, 2004; Robson, 2005). Abundant literature identifies destination image maturity with mediated events (Jago et al., 2003; Lury, 2001). Reid and Ritchie (2011) say that safety in event management is the major topic to be covered by researchers. Risk management plans resulted from the previous belief hoteliers had of safety and security issues. Event managers are often concerned by physical and safety risk where the life of attendees was in danger while financial and other organisational secondary risks can be minimised. These specialists suggest that behaviour is explained by the sum of two factors, subjective norms and attitudes. The degree of control of the attendees is of paramount importance to create a sense of safety in the event venue.

Sports and media events are functional to the preservation of an organic image of destinations. Molloy dwells on the social benefits of festival organisation for the hosting community. She contends that festivals entail a new form of entertainment, in which relaxation and socialisation are promoted in a familiar climate allowing for fun and excitement to predominate. This type of action not only increases the quality of life but also integrates economic and human resources for development. An interesting point of Molloy's work is the role played by the state in event management. Like many other scholars, she accepts that the state intervenes with financial support for the organisation of festivals because it accrues a variety of benefits for interested communities (Molloy, 2002).

Sociological characteristics of mediated events are associated with a ceremonial performance based on two main goals:

"1) to show that the utopia as modeled has always been a driving force in the life of society, such that collective memory exhibits traces of continuous aspiration toward it, and 2) to confirm the feasibility of the proposed transformation by investing the wealth of aspirations in a popular program of action". [Dayan and Katz, (1994), p.180]

A media event ranges from the Olympic Games to a burial. To clarify, even if the word 'media' is identified in popular parlance with the use of television, this is not the case. Media refers to every aspects of human life that are public and reinforce the social commitment in day-to-day issues. Dayan and Katz (1994) define 'mediated events' as celebrations which exhibit a significant importance for the community because they not only strengthen the social bond, but also refer to a founding event in which the community ostensibly forged itself. These types of media events may include military parades, independence days, royal weddings, musical events, sports, and festivals.

In addition, one might admit that effects of mediated events are intended to prevent social fragmentation reinforcing the self-esteem of the entire group. That way, moments of instabilities or uncertainty are a prerequisite for the management of events. This happens because the pre-existing cultural values are often manipulated by privileged stakeholders to their own benefit. In this context, the elite are prone to fabricate facts regarding history to give their own, tendentious meaning to events so as to provide a specific discourse that nourishes a sentiment of belonging. Dayan and Katz enumerate seven characteristics of media events, which explain their importance for the social imaginary:

- media events strengths the legitimacy of elites
- these events resolve earlier stages of dispute or conflicts to forge a strong nationhood
- those leaders who take part of these rites embody a mythical archetype that links them to the lineage of founding parents

- media events not only change the boundaries of society for accepting or rejecting minorities, but also provide neologisms and linguistic resources to denote the new situation of celebration
- like a dream, events act as ideological mechanisms to foster legitimacy of leaders over their citizens.

Dayan and Katz (1994) explain that media events are important to improve the mechanisms of reinforcement and reproduction of societies. Following this, most of the message in these spectacles corresponds with the legitimating of elites and assurance of a broader national being. If some rebellious groups want to cause psychological distress in the community, these types of events offer a particularly fertile space to inflict damage to the society.

Howie (2012) acknowledges the fact that the world and economies have changed forever since 9/11. Given the obsession for security in USA, Howie's research shows how the intervening years have changed the interviewees' viewpoints. There is a strong complicity between terrorists and journalism. Howie's 2012 book examines the limitations of extant conceptual frameworks, and in parallel to other studies, connects the theories of terrorism with late modernity. Howie says terrorism should be defined as more than a political technique or strategy. Instead, terrorism lies in the witness's terror. Terrorism is designed to cause psychological fear, not to obliterate populous cities. In this it contrasts with the mass warfare of the Second World War or the potential war played out over the forty years of the Cold War. Globalisation plays a pervasive role by homogenising the domination of technology, economic linkages, and rationalisation; while on another hand, it entails a process of re-territorialisation based on a bipolar logic of inclusion/exclusion. At the present time, tourism triggers the worldwide displacement of the global system's centre and its recently erected walls and barriers to prevent the crossing of peripheral workers. Poverty and lack of perspective pave the ways for the upsurge of national movements. Even though terrorism can be part of the social discontent that led people to resentment, some groups as in Spain or Ireland opt to participate in politics to gain constitutional representation in the parliament (Del Bufalo, 2002). It is important to note the connection between terrorism, violence, and the law. Whenever some groups are pushed to peripheral conditions, or restricted in political life, they revert to violence as a form of persuasion. When this happens, political discourse sets the pace toward violence (Connolly, 1994). Political terrorism engenders a subtle but powerful message, which in this section we decipher.

5 LEISURE AND TERRORISTS

As the incident of the 15 April 2013 Boston marathon bombing shows, terrorists need public attention. This attention is instrumentalised to achieve their goals, regardless of human safety. Because leisure spaces, especially where mega events occur, are of paramount importance for gaining attention, many terrorists try to accelerate negotiation with states by means of violence in those venues. Munné and Huizinga have demonstrated that leisure is enrooted in the social fabric of society. Leisure, according to Munné, is an ideological instrument of indoctrination that makes sense of the environment (Munée, 1999). Huizinga (1968) emphasises the agonistic nature of competence to define the importance of leisure in the modern world.*Hommo-Ludens,* one of his most famous works, discusses the role of play in the society as a catalyst towards social bonds. Culture is generated by the imposition of play. Following this argument, Huizinga argues that the act of play encompasses five characteristics:

a play surfaces from the ordered freedom

b play emulates an outstanding archetype that goes beyond the normalcy

c in opposition to the ordinary, play is framed in a specific time and space

d although it seems to be chaotic, play exhibits the need of rules and order

e play is subject to non-monetary goals.

Linguistically, Huizinga says that play is excluded from the logic of labour, which situates it in the leisure sphere, and it also enables the principle of civilisation (Huizinga, 1968). The sphere of leisure does apply specific rules, which if broken undermine social trust. Furthermore, leisure opens the doors for uncertainty and ambiguity. Dunning and Elias explain that war and sport represent two different kinds of conflicts that alternate subordination, independence, cooperation, and loyalty. As a vehicle towards pleasure or displeasure, both war and sport connote a combination of rational and irrational behaviour. Furthermore, the existence of contrasting ideologies suggests that sport works as a substitute for war by decreasing the brutality of players. The progress of civilisation depends on the degree to which violence can be disciplined. The civilising process of West is based on three combined factors:

a centralisation of control in a much broader administrative apparatus

b the monopoly of violence exerted by the state alone to foster the security and well being of a community c the ability to introduce in members the social values that keep the society functioning.

Although the state keeps a monopoly of violence by means of technology, and police, the external frontier plays a pivotal role in the process of identity. Leisure and sports seem to be efficient instruments, not only to sublimate the individual violence into acceptable levels, but also to alleviate the daily burden created by the production-led logic of modern capitalism. The deprived self is dangerous to itself and for others. Therefore, modern society tries to control the degree of deprivation its citizens face. The quest for excitement that characterises the world of leisure appeals to a deep-seated discourse so that the self can be duly regulated. In the spaces of leisure, people express their most profound emotions, which remain repressed in normal conditions. What is important to debate respecting to the leisure seems to be the propensity of society to weaken its dispositive control so that emotions may flourish. In doing so, it is not surprising to see that leisure sites are characterised by controlled violence and conflict. But as Elias and Dunning put it, under some conditions this controlled status changes to a state of crisis. This seems to be the reason why not only terrorist-led attacks select international mega-events as primary targets, but also how riots or revolution are initiated in leisure spaces (Elias, 1992a, 1992b; Elias and Dunning, 1992a, 1992b).

Leisure plays a pervasive role because it follows the expansion of empires in the world but at the same time, it weakens the process of control in urban cities. Gomes and Elizalde (2012) introduce in the discussion the idea that leisure does not represent an attempt at recreation or a space of free time. Leisure rather, corresponds with the creative view of changing the environment. Leisure spaces are determined by an encounter between two contrasting forces, which appeal to a state of tension, in which order and disorder converge. In this regard, Igarza (2008) acknowledges that leisure adapts to the existent technology, which leads people to be controlled by technology's structure, but at the same time it entails a change. In sharp contrast with other eras of mankind, the boundaries of leisure and work are gradually becoming blurred. To date, digital technological advances have facilitated faster and more secure access to a broader, interconnected world of information outside of the traditional nation-state control. The question as to whether people can enter a virtual reality and displace their bodies without physical movement became one of the more exciting issues of modernity that social science is just beginning to examine. In another respect, if the term globalisation seems to be a hard concept to grasp even for scholars, the fact is that a growing number of consumers take advantages of digital leisure at work or in schools. Consequently, work time is suffering a radical shift based on a tendency to leisure breaks during work. No less than four decades ago, workers spent their time in one duration, and only stopped once or twice in the day for a recreation for few minutes. Igarza argues that the same technologies people use for working purposes in an office are adaptable for recreational needs. Paradoxically, a process of urbanisation has generated megacities, while at the same time it allows a disaggregation of impersonal persons who increasingly communicate through digital mass consumption, as in networked workers.

Modern cities are characterised by a merger of multiculturalism and cosmopolitanism wherein passers-by often walk or cross avenues rushed by their duties living in the core of vertigo. This hectic style of life seems to be in the quest for more exciting experiences irrespective of the source consumers follow for their recreational needs. The increasing complexity of labour leads workers to an overburdened atmosphere. Recreation and leisure play a crucial role in preserving and refreshing the mind of worker. Recreationists need to escape from rules and state control. This seems the reason why states fail to prevent terrorist attacks in the context of festivals. As noted in earlier sections, terrorism does not obliterate cities or kill thousands of people; they want to encourage a situation of extreme fear alone. Therefore, it is important to understand the psychological effects of terrorism in Western societies.

5.1 Towards a new theory in terrorism prevention policies

It is important not to overlook that not all risks are the same or have the same psychological effects in society. While some risks are often trivialised even though their potential effect would be apocalyptical, others are exaggerated. Risk perception theory explains that the frames of risk are manifold. The following risks are enumerated as the most important studied by literature as factors that may destabilise systems:

a unknown virus and pandemics expanded by international travels

b food safety

c terrorist attacks against civilians

d airplanes or road accidents

e natural or made-man disasters

f crimes, theft against tourists

g political upheaval or instability.

Although each one of the subtypes of risk engenders diverse effects on the tourist mind, we will only deal with terrorist-related acts. To measure the impact of terrorism, we have to evaluate two relevant aspects: the probability of damage signals and the possibility of real harm that a person or the community face in a certain period of time. The effect on social imagination explores the cognitive construal of the risk. Here, conceptual variables are more important for understanding the selectivity of risk perception for a tourist. While many Western tourists trivialise local crime in context of mega events, or attacks produced by the abuse of alcohol, other major risks that seem less probable are emphasised as factors which influence in their destination choice. This seems to indicate that the effects of perception are rooted in the communication process. Sometimes, how these effects are handled are more important than real damage. Sunstein (2006) explains that one of the problems of states is to follow the populist demands of its citizens. First and foremost, citizens are subject to biased impressions of reality that exclude the objective diagnoses of scientists. Those states which claim to cover all safety claims of citizens are involved in a spiral of fear mongering. Lay people misdiagnose the real reason for disasters, including terrorism. They seem to be influenced by the effect alone, and, of course, by the fabrication of news by the media. At the same time, some crimes are considered disgusting, when they are perpetrated on vulnerable targets as women or children, while others are accepted, as when a police officer shoots a delinquent. Policy makers in mega events pay attention to the psychological impacts of risk in tourism and hospitality. Whenever civilians are the targets, public opinion reacts to the attacks, placing their government under the lens of scrutiny.

The psychological impact of terrorism in the media and on the audience is based on three points:

a control threshold

b probability of repetition

c the status of the victims.

Table 1 Psychological impact of terrorism in leisure context

Control threshold	*Probability of repetition*	*Targeted victims*	*Impact*
Low	High	Tourist	High
High	Low	Tourist	Low
Low	Low	Tourist	High
High	High	Tourist	Low

The control threshold refers to the ability of the system to predict the time and place of the attack. The probability of repetition measures the mobilisation of resources to prevent traumatic events. The probability of repetition is proportionate to the effects of events that cannot be avoided. Finally, whenever victims are valorised by their high social status, the impact on social imagination is greater than cases where low status people are affected. This explains, for example, why some few Western tourists have more value than thousands of Palestinians. Terrorists not only recognise this, but also look to attack travellers to have a greater impact on First World states with an immediate decline for the tourism industry. The degree of dependency between tourism and economies in Third World is of paramount importance to understand why they are targets of violence. Table 1 exemplifies and summaries of our argument based on tourism and hospitality only.

Nationality is a variable that correlates with the amount of fear. If tourists are American or British, representing the Anglo alliance, the impact is greater than if victims are Argentines or Brazilians. Also it matters whether the victims are tourists or local residents. For example, when the control threshold is high as well as its probability of repetition, the impact is paradoxically low simply through a process of normalisation. For those events that are constantly repeated, the audience becomes inured to the event and its original impact is undermined. Therefore, natural disasters are more fearful than terrorist attacks. The first combination shows how the impact turns high when the attacks on tourists not only keep a low control threshold – i.e., when the state is unable to prevent the act – and the probability of repetition remains high, suggesting that the state does not have the ability or technology to prevent the next attack. The impact is low if the state develops sufficient resources to control the situation, reducing the odds of repetition. If both the control threshold and repetition probability are low, the impact is high. No matter the probability of repeat, if people think the state is unable to protect then, the process of political instability starts. Last but not least, if the control threshold and repetition probability are high, then the impact changes to low, because the terrorist attacks are normalised. This last point pertains to extent that the public lacks a sensibility of violence and empathy for the suffering of others by media exposure. To some extent, this explains the complicity between media and terrorists. The former domesticates the political instability by over exposure while the latter needs media coverage to transcend the boundaries of state. What remains, is that both feed logic of hate and conflicts. While Western states claim the right of intervention in case of potential terrorist acts, terrorists build a discourse of resentment based on the state's reaction.

6 CONCLUSIONS

In the present essay-review we have discussed the limits of terrorism, its logic, as well as how attacks are planned. Leisure spaces in the global geography are fertile grounds for terrorist acts. Mega-events and festivals exhibit a much broader impact in the social imagination of any community. Affecting the meaning of mega-events, by imposing violent attacks, is a way of creating instability. Unfortunately, the extant safety policies in event management are unable to detect the differences of risks, or to evaluate a clear diagnosis of the messages terrorists generate when they attack civilians. The event management and hospitality industry will reformulate in the next few years policies to help to protect civilians, understanding first that the needs of terrorists are linked to the media. We have provided here a coherent and all-encompassing view as to why terrorists select the mega-events as the primary target of their onslaughts. In the following we summarise the characteristics of terrorism which uses tourism to cause damage to the state:

a Terrorists tend to speculate in order to cause the major impacts on popular sensibilities.

b Hospitality and tourism act as a fertile source to entice terrorists' attention.

c Vulnerable victims receive the attention of the media.

d The state loses its legitimacy when the most vulnerable actors are attacked. This happens simply because the message is: nobody can protect you.

e Leisure and tourist-friendly climates weaken the rules.

f Discrediting the state is one of the most efficient messages of terrorism.

g Whenever the goals of terrorists outweigh the means, the life of people has no worth.

h It is very difficult for the state to regulate behaviour in context of leisure and festivals.

From our point of view, terrorism is a dialectical process between opposing interests which uses violence to gain its ends. This dialectic is not exclusive to a group or a state, but the connection between both. Terrorism has elements or primary aspects that define its existence:

a speculation

b violence

c attacking vulnerable non-combatants

d compliance with the media or journalism

e surprise factor.

Skoll (2007) explains that the function of the state is to maintain the hierarchical status quo by exerting power and violence over populations. In times of low conflict, the legitimacy of the state rests on the market which confers certain stability. In the context of relative chaos and disorder, the state resorts to violence to refashion the threatened order. Similarly, the market mediates among human beings by imposing a state of gratification in lieu of constraints, but the moment the control weakens, fear replaces gratification as motivator to legitimise the ruling order (Skoll, 2007). Terrorism, however, operates as a mirror of Western style of life. Terrorists are not ill-minded persons who want to destroy our civilisation, but a radical experiment of connecting with otherness. Terrorists, let us remind the readers, were educated in the best Western universities, they earned a degree in management or even a PhD. What they learned from the West is how to use leisure to negotiate based on violence and speculation. What many terrorism analysts ignored is that theories of management and marketing studies have facilitated the plans of al Qaeda in synchronising the 9/11 or Boston 2013 attacks.

AUTHOR DETAILS

Geoffrey R. Skoll is an Associate Professor of Criminal Justice at Buffalo State College, State University of New York. His research area includes terrorism and security, postmodernity, and cultural sociology. His recent publications include two books, *Contemporary Criminology* and *Social Theory of Fear*, both published by Palgrave Macmillan.

Maximiliano E. Korstanje is an Associate Professor of University of Palermo and the Editor in Chief of*International Journal of Safety and Security in Tourism/Hospitality*. His research areas are risk perception and disaster-management. With more than 300 published papers and 15 books, he works as editorial board member of more than 20 tourism and disaster-related journals and international councils dedicated to the studies of risk.

REFERENCES

1 Abdel-Azim, T.S. (2010). 'The relationship between the perception of risk and the decision making process of travel of French tourists: the case of Egypt'. Tourismos. 5, 2, 29-47

2 Atkins, B. , et al. (2003). 'The impact of terrorism on tourism and hospitality business: an online debate by experts in the field'. Tourism and Hospitality Research. 4, 3, 262-267

3 Baker, K. , Coulter, A. (2007). 'Terrorism & tourism: the vulnerability of beach vendors livelihoods in Bali'.Journal of Sustainable Tourism. 15, 3, 249-266

4 Berlang-Adell, M.J. (2004). 'Turismo y Poder: las transformaciones de una fiesta popular en Marruecos'.Pasos: Journal of Tourism and Cultural Heritage. 2, 1, 25-45

5 Bianchi, R. (2007). 'Tourism and the globalization of fear: analyzing the politics of risk and (in)security in global travel'. Tourism and Hospitality Research. 7, 1, 64-74

6 Castaño, J.M. (2005). Psicología Social de los Viajes y el Turismo. Madrid:Thomson Ed.

7 Cohen, R. (1990). Culture and Conflict in Egyptian-Israeli Relations: A Dialogue of the Deaf.Bloomington:Indiana University Press

8 Connolly, W. (1993). The Terms of Political Discourse. 1st ed., Princeton:Princeton University Press

9 Cooper, H.H.A. (2001). 'Terrorism: the problem of definition revisited'. American Behavioral Scientist. 44, 6,881-893

10 Crenshaw, M. (1983). 'Introduction: reflections on the effects of terrorism'. Terrorism, Legitimacy, and Power.Middletown CT:Wesleyan University Press , 1-37

11 Crenshaw, M. (1995). 'Thoughts on relating terrorism to historical contexts'. Terrorism in Context. University Park PA:Pennsylvania State University Press , 3-24

12 Crompton, J. , Mckay, S. (1997). 'Motives of visitors attending festival events'. Annals of Tourism Research.24, 2, 425-439

13 Dayan, D. , Katz, E. (1994). Media Events. The Live Broadcasting of History. Cambridge:Harvard University Press

14 Del Bufalo, E. , Ceceña, A. , Sader, E. (2002). 'La Restructuración neoliberal y la globalización'. Guerra Infinita: hegemonía y terror mundial. Buenos Aires:CLACSO , 39-62

15 Elias, N. , Elias, N. , Dunning, E. (1992a). 'Introduction'. Sport and Leisure in the Civilizing Process. Buenos Aires:Fondo de Cultura Económica , 30-81

16 Elias, N. , Elias, N. , Dunning, E. (1992b). 'Genesis of sports as a sociological issue'. Sport and Leisure in the Civilizing Process. Buenos Aires:Fondo de Cultura Económica , 157-184

17 Elias, N. , Dunning, E. (1992a). 'The quest for excitement in leisure fields'. Sport and Leisure in the Civilizing Process. Buenos Aires:Fondo de Cultura Económica , 83-115

18 Elias, N. , Dunning, E. (1992b). 'Leisure under spectrum of free-time'. Sport and Leisure in the Civilizing Process. Buenos Aires:Fondo de Cultura Económica , 117-156

19 Floyd, M. , Pennington-Gray, L. (2004). 'Profiling risk: perception of tourist'. Annals of Tourism Research. 31, 4, 1051-1054

20 Gomes, C. , Elizalde, R. (2012). Horizontes Latinoamericanos del Ocio. Belo Horizonte:Universidade Federal de Minas Gerais , Editorial

21 Hall, M. (2002). 'Travel safety, terrorism and the media the significance circle of the issue attention cycle'.Current Issues in Tourism. 5, 5, 458-466

22 Henderson, J.C. (2008). 'Managing crises: UK civil aviation, BAA airports and the August 2006 terrorist threat'. Tourism and Hospitality Research. 8, 2, 125-136

23 Howie, L. (2012). Witness to Terror: Understanding the Meaning and Consequence of Terrorism. New York:Palgrave

24 Huizinga, J. (1968). Hommo Ludens. Bogota:Emece

25 Igarza, R. (2008). Leisure's Bubbles: A New Form of Cultural Consumption. Buenos Aires:La Crujia

26 Jago, L. , et al. (2003). 'Building events into destination branding: insights experts'. Event Management. 8, 1,3-14

27 Korstanje, M. (2009). 'Re-visiting the risk perception theory in the contexts of travels'. ERTR: E-review of Tourism Research. 7, 4, 68-81

28 Kozak, M , Crotts, J. , Law, R. (2007). 'The impact of the pereption of risk on international Travellers'.International Journal of Tourism Research. 9, 4, 233-242

29 Kucukaltan, D. (2006). Tourism and Terrorism: an Experience of Turkey and the World. Nebraska:Universe

30 Kuto, B. , Groves, J. (2004). 'The effects of terrorism: evaluating Kenya's tourism crisis'. E-review of Tourism Research. 2, 4, USA:Texas A&M University , 88-95, Pero ¿como definir un acto 'terrorista?

31 Lepp, A. , Gibson, H. (2008). 'Sensation seeking and tourism: tourist role, perception of risk and destination choice'. Tourism Management. 29, 1, 740-750

32 Liang, Y. , Illum, S. , Cole, T. (2008). 'Benefits received and behavioral intentions of festivals visitors in relation to distance travelled and their origin'. International Journal of Event Management Research. 4, 1, 12-23

33 Lury, G. (2001). 'Thoughts on decision branding, British bulldog days are numbered'. Brand Strategy. 1-3

34 Molloy, J. (2002). 'Regional festivals: a look at community support, the isolation factor and funding sources'.The Journal of Tourism Studies. 13, 2, 2-15

35 Moten, R.A. (2010). 'Understanding terrorism: contested concept, conflicting perspectives and shattering consequences'. Intellectual Discourse. 18, 1, 35-63

36 Munée, F. (1999). Psicosociología del Tiempo Libre. México:Editorial Trillas

37 Peattie, S. , Clarke, P. , Peattie, K. (2005). 'Risk and responsibility in tourism: promoting sun-safety'. Tourism Management. 26, 3, 399-408

38 Prentice, R. , Andersen, V. (2003). 'Festivals as a creative destination'. Annals of Tourism Research. 30, 1, 7-30

39 Reid, S , Ritchie, B. (2011). 'Risk management, event managers attitudes, beliefs and perceived constraints'.Event Management. 15, 4, 329-342

40 Ritchie, B. , Sanders, D. , Mules, T. (2007). 'Televised events: shaping destination image and perception of capital cities from the couch'. International Journal of Event Management Research. 3, 2, 12-23

41 Robson, L. (2005). 'Risk management for meetings and events'. Annals of Tourism Research. 35, 3, 840-842

42 Skoll, G. (2007). 'Meaning of terrorism'. International Journal for The Semiotics of Law. 20, 1, 107-127

43 Steiner, C. , Scott, N. , Jafari, J. (2010). 'Impacts of September 11: a two sided neighborhood effect?'.Tourism and Muslim World. West Yorkshire:Emerald Group Publishing , 181-202

44 Sunstein, C. (2006). Riesgo y Razón: seguridad, ley y medioambiente. Buenos Aires:Editorial Katz

45 Tarlow, P. , Korstanje, M. (2011). 'Tourism risk management in the age of terrorism'. Economía Autonoma. 7, 1, 25-40

46 Task Force on Disorders and Terrorism (1976). Disorders and Terrorism. Washington DC:National Advisory Committee on Criminal Justice Standards

47 Vukonic, B. , Scott, N. , Jafari, J. (2010). 'Do we always understand each other?'. Tcurism and Muslim World.West Yorkshire: Emerald Group Publishing , 31-45

48 West, B. (2008). 'Collective memory and crisis: the 2002 Bali bombing, national heroic archetypes and the counter-narrative of cosmopolitan nationalism'. Journal of Sociology. 44, 4, 337-353

CHAPTER 4

Sport events and human rights: positive promotion or negative erosion?

Andrew Adams[a*] & Mark Piekarz[b]

[a] Centre for Event and Sport Research, Bournemouth University, Poole, UK

[b] Institute of Sport & Exercise Science, Worcester University, Worcester, UK

ABSTRACT

In this paper, we build upon recent scholarship on sport event legacies to identify, categorise, and describe the key processes underpinning sport event interactions with human rights (HR). It develops a simple, representative model to illustrate the points where sport events bisect with HR and considers what factors can modify these impacts. The development of this model is based on a meta-review of literature and examination of case studies. It is clear from our analysis that sport events are malleable, symbolic, and political occurrences that can be positioned to provide evidence and support of benefice or harm to the cause of HR. The model also provides a nuanced approach to consider how sport event organisers may begin to think about the tactics and strategies that might be employed, and how they might leverage HR through their sport event. The model also indicates that HR, being similarly malleable political tools, are paradoxical in application in the sport event context and consequently cannot be assumed to be taken-for-granted as event outcomes or outputs.

KEYWORDS

human rights, sport events, sport event impacts, mechanisms and processes

1. Introduction

Can sport events improve the human condition and enhance people's human rights (HR)? It is a pertinent question given the greater attention being paid to the outcomes of sports events. This reflects changes in social and political attitudes; growing scrutiny by pressure groups and the media; and the increasing necessity to justify spending vast amounts of money in order to stage sport events. The bold rhetoric of the transformative power of sport to do good is a common theme in event bidding processes. Lord Coe's 2005 Olympic bid speech spoke of reaching young people, inspiring them to participate and celebrate humanity (BBC, 2014); the Chinese government, in 2001, on winning the right to stage the Olympic games, declared unequivocally how the event would enhance 'all social sectors – education, medical care and human rights' (cited by Booth, Kidd, & Toohey, 2001). The list could go on. Indeed, articulations of this sort, and the importance of these statements is perhaps encapsulated by Nelson Mandela's often cited speech about the power of sport for good (Laureus Sport for good Foundation, Foundation overview 2013), Mandela said:

Sport has the power to change the world. It has the power to inspire. It has the power to unite people in a way that little else does. It speaks to youth in a language they understand. Sport can create hope where once there was only despair. (Monaco, 2000, Laureus lifetime achievement award)

Yet caution is needed about these 'go to' arguments of the benefits of sport. Statements about the value of sport are not evidence that the many stated benefits actually exist or accrue. For this paper they represent part of the rhetoric of sport event justification. Essentially these sport evangelical arguments loosely promote a mythopoeic view of sport (Coalter, 2007), where sport is conceived of as a panacea, a self-evidently good thing (Rowe, 2005), virtuous (Grix & Carmichael, 2012), morally good, and promoting pro-social behaviour (see Coakley, 2011 for a critical view). The rhetoric of many key sport event stakeholders often normalises and then naturalises the idea that sport events will enable transformative outcomes as part of the logic of hosting such events. Few academics would accept such an uncritical position, and there is a growing literature that is more concerned with interrogating and scrutinising evidence and sources of evidence to establish what works, for whom and in what circumstances (Adams & Harris, 2014; Coalter, 2007).

It is clear therefore that a sharper picture is needed to make any assessment of the impacts on HR. This is important for a number of reasons. First, because HR are often overused as a rhetorical device, taken for granted but overlooked or omitted as other 'real-world' costs impinge on sport event organisers. Second, because many interpretations and theories of HR suffer from what Sen (2004, p. 315) has called a 'mushiness' or softness of conceptual grounding, reflecting a fluidity to the language of HR (Cmiel, 2004). Third, because the discussion about sport events and HR has become entangled with the concept of corporate social responsibility (CSR), which Dowling, Robinson, and Washington (2013, p. 269) define as companies integrating 'social and environmental concerns in their business operations and in their interactions with their stakeholders on a voluntary basis'. Many national sport organisations (NSOs) and international sport organisations (ISOs) such as the International Olympic Committee (IOC) and the Federation International Football Association (FIFA) have certainly engaged with the language of CSR led by ethical and/or business demands. Finally, the gap between the claimed rhetorical benefits and the actual realities of sport events has been an issue that an increasing number of pressure groups, journalists, and academics have scrutinised, often creating 'uncomfortable truths' for those leading the sport and the events.

This last is crucial and has a number of sub-issues. The first concerns how critical agents may make reference to sport event rhetoric, charters, and plans that outline the importance of ethical behaviour and respect for HR, in order to hold the event organisers to account. For example, the Centre of Housing Rights and Evictions (COHRE), in their report on the displacement impacts of the Beijing Olympic Games (2008), issued clear statements to the IOC to abide by their own principals and code of ethics enshrined in the Olympic Charter, which embody many key principals about HR. The second centres on Wagner's observation about the 'displacement of scope' (1964, cited by Coalter, 2013, p. 3), which describes the process of wrongly generalising the potential micro effects of sport activities, to the meso and the macro level. Put simply, because *there is*evidence of benefits accruing from the playing of sport, it does not mean that the same benefits will transfer to a sport event. These issues are critical as they underscore the need for a model that clarifies the processes, interactions, and system features that condition any HR–sport event relationship.

Contextually, this paper should be considered within the much broader field that incorporates sport event impact analysis and leveraging. The focus on HR reflects the increased discussion of rights used in sport event processes, whether as part of the rhetoric for event justification, or by pressure groups using events to expose broader issues about HR. To assist and enhance the development of such a discourse, this paper identifies a model outlining *how*, *why*, and *where* sport events impact on HR, as well as highlighting the factors that can modify these impacts. The model also indicates the mechanisms and/or levers that act to positively push or negatively pull on HR.

2. SPORT AND HR: MEANINGS AND LIMITATIONS

HR are primarily (according to Sen, 2002, 2004) ethical demands that can and do inspire legislation, generating reasons for agents to act and can create a range of obligations among those agents who are able to act, as understood in Western liberal traditions (Booth, 1999, Griffin, 2008, Sen, 2004 etc.). In taking the universalist position and accepting the 1948 UN declaration as the cornerstone of our analytical framework, the notion of human dignity or personhood (Griffin, 2008), which is commonly thought of as an attribute of humans (Valentini, 2012), is fundamental to considering HR discourse. In this respect HR are not just abstract values, but are indicative of a set of social practices to realise these values – social practices that when embedded in sport events can and should enable particular outcomes via a range of leveraging strategies.

The UN declaration of HR in 1948, like previous assertions of man's rights, emerged from a period of barbarism and was founded on a liberal and universalist position against which the majority of Western nations sought to judge primarily other nations in terms of how free, safe and civilised they were (Brown, 1999, Griffin, 2008, Sen, 2004). In short the Universalist position is one that makes ethical demands based upon a shared common humanity that confers dignity and personhood on individuals. The dominance of the Universalist conception of HR is also not without controversy concerning charges of cultural imperialism and cultural relativism. The automatic assumption that HRs are universal is trumped, by relativists, with arguments that place the social construction of the individual and the historical and cultural location of those HR to the fore.

A common relativist trope is that the universality of HR acts as an 'ideology which is a cover for the imposition of western values' (Booth, 1999, p. 52). It is clear that in the sport domain ongoing globalisation processes have brought attention to the applicability value and meaning of HR. Giullianotti, (2004, p. 362), for example, asks the question 'Is it possible that "human rights" is an essentially Western cultural conception that, in truth, carries no universal reach?' Post-colonialist theorists have argued that sport as an institution provides a dominating axis along which Western cultural mores have been passed and that HR in general, but certainly in the field of sport, reinforces dominant Western hegemonic notions of morality and reasserts power inequalities through its practice. The purpose here is not to attempt to reconcile these problems in the context of sport (see Giulianotti, 2004 who does attempt to), rather we wish to sensitise the reader to some of the controversies inherent in discussing and applying notions of HR in any socio-political context.

The rationale for basing our modelling on a universalistic framework is primarily one of pragmatic political applicability in the context of sport events; secondly, we concur with Giulianotti's (2004, p. 358) argument that the 'cornerstone' of any discussion of HR must be the

Universal Declaration. Thirdly, for all its faults and criticisms, the document has some remarkable qualities in terms of the directness of the language and its brevity. Finally, we also agree with Sugden (2010, p. 268) who notes, that the declaration 'is one of the few touchstones for governing activism that has near universal approval'.

One difficulty is how all of the rights, articulated in the convention and many subsequent charters, codes, and legislations, are represented in a model which can then be used to gauge the pull or push impacts of sport events? Without appearing to be reductivist, but for purposes of clarity, the 30 articles of the 1948 Declaration are represented as coalescing around five key themes, which are summarised in Table 1. The themes reflect a universalistic HR discourse that has facilitated communication and interaction around the world and is important if any model based on these themes is to have global purchase in application (Cmiel, 2004).

Table 1. Five themes of HR.

Theme	Focus	Articles
Freedom	of movement, association, expression and organising governments	1,13, 14, 21, 19, 17 and 20
Protection	from the state in relation to torture, detention and expropriation of property	4, 5, 9, 10, 11, 12 and 14
Access	to an impartial legal system with the presumption of innocence	6,7,8 and 9
Equality	treatment for all members of the community	1, 2, 3, ,6, 7, 8, 10 and 11
Ability	to maintain their physical, social and psychological well-being through work, access to welfare services, education and not exposed to dangers	16, 18, 20, 22, 23, 24, 25, 26 and 27

Source: 1948 UN Universal declaration of human rights.

3. SPORT, SPORT EVENTS AND HR

The 1948 Declaration did not specifically declare sport as a right, but can and should be regarded as an implicit component of various rights, such as those relating to freedoms of cultural expression, association, and leisure (articles 24 and 26 are particularly relevant in this respect). Since the 1948 declaration, various subsequent declarations and acts, such as the Brighton Declaration (1994), have attempted to clarify both sport as a right in itself and how it relates to certain groups, such as women or children. Today, it is possible for sport to be presented as a tool or mechanism that can impact on HR; as being a right in itself; as a means to enhance quality of life, promote peace; and deliver and improve people's HR.

This seemingly continual expansion of what might constitute HR has been readily advocated by many national and ISOs, not least because it helps in global promotions of sport ideologies and perhaps cements two 'lingua francae of the globalised world' (Cmiel, 2004, p. 126) as hybridised and rhetorically powerful language tools to promote dominant notions of the power of sport (Hartmann & Kwauk, 2011). For example, the IOC Charter (2014, p. 11) states as its fourth principle that:

The practice of sport is a human right. Every individual must have the possibility of practising sport, without discrimination of any kind ... in ... a spirit of friendship, solidarity and fair play.

Similarly, the United Nations Sport Development and Peace (UNSDP) International Working Group has identified sport as a 'fundamental right for all' (2014); indicating that sport can be a powerful tool with which to further develop HR, particularly when sport activities remove barriers, reduce inequalities, and promote peace. From those examining histories of HR (Angle,2002, Cmiel, 2004, Hunt, 1996, Lauren 2011, Sen, 2004), it is apparent that rights are malleable and can and do change over time, and despite the universalistic logic applied throughout this paper, that in practice HR are culturally and socially bounded. It is clear therefore that what is, and is not accepted, as a HR has changed over time, and that internationally, understandings of HR have expanded to specifically include sport. It is in this context that sport, representing a universal HR idiom, can acquire local meanings that are contested and develop over time. It is also vital to point out that because sport can be framed as a right, there is not a natural corollary that by people playing sports or staging events that HR are enhanced or protected. The two are separate. To mix them up is to displace the scope. Sport events can involve the playing of sport, but they also impact, bisect, and interact with people in more complex ways which needs a more intimate and precise analysis

Many (e.g. Booth et al., 2001, Donnelly, 2008, Giulianotti, 2006, Kidd & Donnelly, 2000, Lenskyj, 2006, Sugden, 2010) have given varied insights into the sport/HR relationship, along with invaluable critiques outlining the relative weakness of sport in improving HR. The dominant discourse of these, and other critiques before them, is that sport is a relatively weak force and/or mechanism in helping to bring about political change; change, which might have the corollary of improving HR. Interestingly, many of the examples cited by these various commentators provide instances of where sport, acting in a variety of different contexts, appears to erode, or pull away people's rights. Additional criticisms, concerning claims made about sport, have been levelled by a range of observers. Coalter (2007) has been consistent in arguing forcefully that many of the unrealistic outcomes attributed to sport are founded on mythopoeic interpretations of sport that have been promoted by a range of sport evangelists (Coakley, 2011). Furthermore, Collins (2010) and Grix and Carmichael (2012) point to a cult of sport that undermines participation policies and systems and Levermore who points to the 'paucity' of monitoring and evaluation in establishing whether claimed outcomes match actual outcomes leaving some convinced that the use of sport 'inevitably leads to positive outcomes'

(Levermore, 2011, p. 341). Adams and Harris (2014) note the increased need for more 'appropriate evidence' in line with the demands of '…accountable organisations to demonstrate effectiveness and impact' (p.140). Kay (2009) is unequivocal on this matter, arguing that 'the claimed benefits attributed to sport overreach the research base as the evidence of sports social impacts is unsatisfactory' (p. 1178).

4. A REPRESENTATIVE MODEL FOR ANALYSIS: POSITIVE PUSH OR NEGATIVE PULL?

The model presented here complements other frameworks, such as Gratton and Preuss's (2008) three-dimensional legacy cube, which is based on the; positive or negative impacts; planned and unplanned impacts; and the tangible and intangible impacts. The representative model we develop is articulated by the legacy cube in that the focus on HR gives a specific area of consideration in relation to how those HR can be positively or negatively impacted on. It is designed to account for both the planned and unplanned impact and provides for a more tangible measurement of sport event impact on HR. In this sense it is a model which has practical applications, designed to assess tangibility and provide measures to help test some of the claims about improving rights often made about sport events.

The model adopts a Universalist position as an underpinning foundation. Simply put this means that HR and any improvements of, or to, them is considered as something positive, whilst situations where HR are eroded are seen as negative. Such an approach means that the model can suffer many of the more generic criticisms and limitations made about rights discussed earlier. In response, it is because of the dominance of the Universalist rhetoric used in sport event justification, that this model apes the language: in short, if organisations seek to use the language of HR enhancement to help justify sport events, then they should expect to be measured using the same language and standards.

In Figure 1, a model is developed to represent how a sport event can interact with HR. One of the features is to use the imagery that the words 'push' and 'pull' create, when looking at the potential impacts. It is this suggestion of movement, which is helpful in conceptualising our model that helps in testing the claims made by many sport event organisers that sport events can help improve and enhance HR. Certainly this language is useful in attempting to make sense of the tangibility of sports events in having an effect on HR. Thus, we term these impacts as positive pushes that promote HR and negative pulls that degrade and/or erode rights. These notions of pulls and pushes are considered in relation to the five key themes, presented in Table 1. The model has three central elements to it: the points of interaction; the factors that can modify the impacts; and the levers and/or mechanisms that help push and promote, or pull and erode HR. The explanation and illustration of these three sections structure the final parts of the discussion in this paper.

5. POINTS OF INTERACTION BETWEEN SPORT EVENTS AND RIGHTS

Gratton and Preuss's (2008) legacy cube helps convey the multiple impacts from sport events which relate not only to the actual sport event, but also to the infrastructure developments and the jobs which are directly and indirectly created from the whole event staging process. Within all of these areas, there are multiple points where the event process bisects with people's lives, which in turn can represent a point where there is the potential for HR to be pushed or pulled. These points of intersection are the tension points between HR and events. It is possible to summarise three key points where events can bisect or interact with people's lives, which in turn creates a potential point of impact on their rights. These are:

- land-use,
- sport event activities, and
- work and employment practices

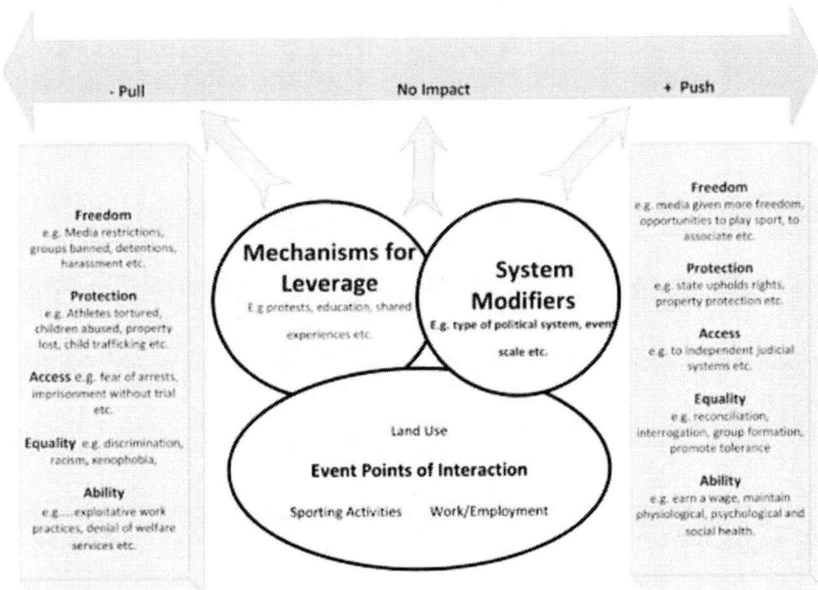

Figure 1. A representative model of sport and human rights points of interaction.

In considering land use, it is the impact of the construction of sporting facilities together with associated infrastructure developments that may impact on people's HR. The construction of roads, hotels, stadiums, and/or the removal of recreational spaces, for example, creates many pressures and potential tension points and claims on land where people live. Whilst many of these developments can enhance physical environments and play a role in improving peoples' quality of life, they can also be a rich vein to explore for finding evidence that the staging of sport events can be detrimental for HR.

Numerous examples can be found to illustrate these points of tensions, where people may be denied *protection* of their property rights, or are unable to *access* a fair hearing to stop developments, together with inequalities being reinforced. Indeed, find any large sporting event, and one can easily find examples of these issues, where individual and community HR are negatively pulled away. For example, in South Africa there were many cases of how people were forcibly evicted from their homes, because of the 2010 Football World Cup, and moved to what were described as temporary relocation areas, which Smith (2010) has described as having more of an air of a prison camp. Similar concerns have been raised by Raquel Rolnik, the UN special Rapporteur on the right to adequate housing, about allegations of displacement and evictions, lack of compensation, and unreasonable speed of action prior to the 2014 football World Cup and the 2016 Rio Olympic Games, respectively (UN 2014).

To balance these issues of HR being eroded, it seems necessary for more nebulous arguments concerning push effects of sports events to be developed. These arguments often concern how infrastructure developments and improvements in the built environment can help enhance the opportunities for leading a life of dignity and/or enhancing the opportunities for cultural expression. For example, Kissoudi (2010) observes how the Athens Olympic Games spurred on investment in various cultural and heritage works, such as improving visitor facilities to archaeological sites. These various works, it can be argued contributed to people's *ability* to express and formulate their cultural identify, with the improvements in the quality of the environment also making a contribution to their sense of well-being, albeit, as Kissoudi argues, only for a short time

In relation to the actual sporting spectacle itself, the impact on HR again may vary. Health, as illustrated earlier, is often considered in the context of HR. Whilst sport events may be considered relatively passive, limiting any individual claim to physical health benefits, it is possible and advisable, however, to consider the social and psychological health benefits. Sport events can be a source of stimulation and happiness, and consequently have the potential to contribute to individual HR to maintain their sense of social and psychological well-being, whilst also creating the *freedom* of opportunity for collective identity and cultural expressions. These positive pushes can of course be countered by examples of negative pulls; the factors which can drive communities to unite and bind, do so at the cost of reinforcing differences, and intensify feelings of 'them' and 'us' divisions. This paradox for Giulianotti (2004) lies in sport's symbolic value to be a source or focus for social conflict and prevent integration of groups in society, which is often expressed through nationalism, sexism, racism, and other forms of xenophobia.

Another point to consider is how the participants themselves have their own rights enhanced or eroded. Elite athletes have the potential to gain many new freedoms and opportunities to enhance the quality of their lives. Yet, it is also possible to raise questions about the restrictions placed on athletes' *freedom.* For example, Schneider (2004) argues that the many restrictions and demands placed by doping authorities (particularly Worlds Anti Doping Agency), such as the restrictions of movement and invasions of privacy and confidentiality negatively erode (or pull) the HR of athletes. Schneider in balancing the argument that sport is different, argues that 'nothing is so special or different that it cannot escape all moral scrutiny from those outside sport' (2004, p. 453). There are also well-documented instances of physical (Kelly & Waddington, 2006; Kerr, 2010; Lang, 2010), emotional (Gervis & Dunn, 2004; Kerr & Stirling, 2008, 2012; Stirling & Kerr, 2007, 2009) and sexual abuse in sport (Brackenridge, 2001; Cense & Brackenridge, 2001; Fasting, Brackenridge, & Kjolberg, 2013; Hartill, 2009).

The final point of tension considers how sport events can directly and indirectly create employment opportunities, both paid and voluntary, the quality of which can have profound implications for a person's HR. One of the criticisms of the principle of HR is that they are intimately tied in with economic development: that they are easy to have and support where there is a degree of economic affluence. Using a simplistic deterministic logic, the argument can be created whereby any jobs which are created are seen as good, which provide money and an opportunity to improve the quality of life, which in turn makes the pursuance and preservation of rights easier.

There are of course problems with such deterministic arguments. One important criticism concerns the quality of the jobs and the degree to which people are protected from exploitative practices. Again, it is relatively easy to find examples of how jobs created via event processes can pull away at people's HR. For example, Donnely and Petherick (2006) note that whilst a manufacturer might have a factory in a developing country that may create certain economic freedoms, the factory itself may use exploitative working practices, which can be both detrimental to workers' health and safety, and place restrictions on their freedoms. The jobs created by the need to construct sport facilities and other infrastructure projects is also an area where pressure groups have been critical about the poor, exploitative, and often dangerous working conditions faced by immigrant workers. Amnesty International have been a vociferous critic of FIFA in highlighting how the use of immigrant labour in Qatar's construction sector is 'rife' with abuse and 'exploitation' in the construction of World Cup facilities. The resulting high number of accidents and deaths represents a damning indicator of exploitative practices, processes, and procedures (Amnesty International, 2013).

6. SYSTEM MODIFIERS

The sport event process will have many points of intersection with people's lives, which creates a point of tension and a potential to push or pull HR. From the literature reviewed, two broad modifiers can be identified which can further modify the impact outcome on rights. These can be summarised as:

- sport event scale and
- type of political system

Considering event scale impact, simply put, the bigger the event, the more people's lives have the potential to be affected, together with the medium- to long-term legacies that people's lives are transformed, for good or bad. Horne (2010, p. 33) argues that mega-events 'provide multiple meanings' for different groups of people, so it is not surprising that positive and/or negative impacts can be found depending on who is being looked at, or even who is doing the looking. Thus, a hallmark event, such as the Olympics, will have many different impacts in comparison to smaller, grass-roots-based sport events. The sheer scale, resources needed, and the number of people whose lives are bisected by the event mean that in terms of HR, differences will be numerable and differentiated according to relative scale.

The second key point to consider is how the nature of the political system can further refine and shape the impacts on rights. Care should be taken about the simplistic use of terms of democratic and non-democratic systems and the relative risk posed to HR. Yes, it is easy to find examples of how undemocratic, authoritarian regimes have used a sport event as a point to reaffirm and retrench power, sometimes with the use of state violence. Commonly cited examples include the 1936 Nazi Olympic Games; the 1968 Mexico Olympic Games; the Argentinian military junta's use of the 1976 football world cup; the 1988 South Korean Olympics; and the 2008 Beijing

Olympics and the issue of China. This litany of large-scale event impact negative pulls presents a number of paradoxical scenarios. For example, in 1968 the Mexican president, Gustavo, Diaz Ordaz, wanted to legitimise his regime and present a positive image of Mexico to the outside world. Ironically, the Games created an opportunity for people to protest about the government, which subsequently was met with a great deal of violence, as hundreds of student protestors were murdered by security forces to secure order. The protests and the brutality of the repression did however lay the seeds for later reform (Giulianotti, 2004). Similar arguments are made by Spaaij (2009) in relation to South Korea, or Hough (2008, p. 1292) in relation to Argentina where he argues that protests actually helped to expose the violence of the regime and its lack of legitimacy.

Whilst democratic systems may not use the same degree of force, violence, or mass arrests to deal with protests groups, this is not to say that they do not use force, or put in place measures which have the potential to pull away HR. Indeed, for both democratic and authoritarian countries pressure groups have pointed out the risk of rights and civil liberties being eroded by issues of security and pressures to protect commercial sponsorship rights. Giulianotti (2004) cites the example of how the Queensland legislature passed legislation which effectively cleared aboriginals from the street in preparation for the 1982 Commonwealth Games in Brisbane; for the London 2012 games, there were criticisms of the 2006 Olympic Act, which gave authorities repressive powers to force entry into private property to remove unauthorised advertising and/or protest banners. Indeed, the issue of event security has driven a huge increase in surveillance and restrictions on freedoms that various pressure groups have raised concerns about (Jennings & Lodge, 2009).

7. LEVERS FOR CHANGE

The third critical part of the model enables consideration of the points of leverage. The notion of leveraging has increasingly been applied to sport events; for social leveraging (Chalip, 2006); for community benefit (Kellett, Hedeb, & Chalip, 2008); and event themed regeneration (Smith & Fox, 2007). This concept deals with the growing criticism and perceived limitations of impact studies. The critique argues that the identification of a potential impact and legacy is not the same as saying it will occur; what matters are the mechanisms or tools, which allow any positive impacts to be 'levered' into place.

In relation to the model presented in this paper, leverage refers to the tools and/or mechanisms that need to be used in order to push and improve HR, or at worst prevent them from being eroded or pulled away. Kidd and Donnelly (2000) note that whilst the state has a pivotal role in enhancing, protecting, and upholding rights, individuals and organisations, such as commercial businesses or governing bodies of sport, are not exempt from these obligations. What this suggests is that there are a variety of levers that governments, businesses, pressure groups, or communities can use, and that these reflect six key rhetorical and tangible themes of sport event management. They are:

- infrastructure developments and facilities;
- protests, riots, strikes and violence;
- symbolism and shared experiences;
- education;
- boycotts, embargoes and diplomatic talks; and
- commercialisation and brand exploitation.

Infrastructure development is discussed earlier, however it is important to note that the reason why it is also cited as a point of leverage is because the type of facilities and infrastructure developments can help shape individuals and a community's quality of life by creating, or removing, opportunities for cultural expression and association.

The second related theme concerns the reality of sport events to create many powerful communication opportunities as well as amplifying messages, which can be made more powerful by levering the symbolism of the event. It is little surprise then that sport events can be as attractive to groups or individuals who wish to protest as they are to sponsors and advertisers. As Lenskyj (2010, p. 15) argues, events can draw the 'eyes of the World' and can expose regimes and organisations abuses of rights, whilst also potentially give a voice to the less powerful, marginal groups, or expose corrupt practices and forcing changes in governments. There are various methods which groups and individuals can use to try and lever these opportunities to communicate messages and to put pressure on governments to change. At the most extreme can be the terrorist group who may target a sport event to focus attention to their cause. Others, ranging from individual athletes staging protests, to groups outside of sport, may use both the build-up to the event and the actual event itself as a focus for protest and demonstration.

Demonstrations and protests at sport events have had mixed success. Amnesty International has increasingly used various sport events, such as the Olympics and Football World Cup to raise HR issues of abuse, ranging from the Chinese governments detention of citizens, to the conditions of foreign workers used to build stadiums in Qatar. Christian Aid (among others) has also raised the issue of child labour in the manufacturing of footballs and sport clothing, and used key sport events, such as the 1998 Football World Cup, to shine a spotlight on the issue. FIFA in response, have followed a pattern common to large transnational stakeholders of: ignoring the issue;

denying the issue; being embarrassed by the issue; reluctantly engaging with the issue; and then positively supporting the issue. After a decade, it is now common to see the very public rhetorical engagement with this issue of child exploitation in manufacturing of sport goods by both sport organisers and commercial companies. Hough (2008) has illustrated how FIFA, in conjunction with the International Labour Organisation (ILO), and supported by sport manufactures such as Puma and Adidas (companies who received a great deal of criticism for their past practices), all now give very public support to the Red Card to Child labour campaign. Similarly, Timms (2012) explores the move for a 'sweat-free' Olympics campaign, which has had many points of success in the run up to the 2012 Olympic Games.

Third, levering points of shared experiences and symbolism is a theme that provides some powerful tools for both good and bad. Kathy Freeman's winning of the 400 metres at the 2000 Sydney Games was used to represent both the divisions in Australia between aboriginal groups, and colonisers, and the process of reconciliation and integration (Bairner & Molner 2010, P. 4). In South Africa, the cricket, rugby, and football world cups have been key events in the process of nation building and reconciliation (Donnelly, 2008, p. 38). The presentation to South Africa at the 1997 Rugby World Cup of the winning trophy by Nelson Mandela wearing a Springbok shirt is a potent example of the power of symbolism and the shared experience. Schulenkorf (2010) examines Sri Lanka to provide some evidence of how sport development and events can create moments of togetherness for ethnic reconciliation, but as he notes, these are modest and need to be done as part of a wider strategy for reconciliation, along with consideration about transference of scope arguments. This shared experience mechanism can also have its negative side. The history of sport events indicates the potential for negative pulls such as the use by Croatia and Yugoslavia, of football events, in the 1990s, to help mobilise violent nationalism, hooliganism, and racism against other ethnic groups (Hough, 2008).

Education can also be a lever to change behaviour, such as encouraging tolerance of other groups. No large sport event goes by without a variety of educational campaigns leading up the event, during the event, and after the event. For example, there are many examples of sport organisations or athletes giving endorsement to campaigns or initiatives which can be intimately entwined with HR, such as FIFA's endorsed campaign of 'kick racism out of football' illustrates. How successful such campaigns are is still an area of contention, with relatively little evidence to support positive or negative outcomes.

Another mechanism used to try and transform governments' actions, policies, or even change can be through external political pressure, if not power, over particular decisions. Boycotts have a mixed history here, ranging from the calls to Boycott the 1936 Olympics, to the boycotts of the 1980 Moscow games by America, and the 1984 LA games by the Soviet Union. Again, the case of South Africa can sometimes be presented where boycotting sport did act as another pressure point for political change. The reinstatement of South Africa back into the Olympics in 1992 in turn helped act as a further symbol or signifier of change (Bairner & Moner, 2010, p. 10).

Commercialisation and brand exploitation provide some more interesting examples, where some have argued that the drive to commercialise interests and brands has eroded HR. For example, Eick (2010), in his examination of the creation of restricted fan zones at the 2006 German World Cup, argues that it was not about giving fans without stadia tickets the opportunity to watch the matches, but about marketing and advertisement rights. Eick argues that these zones were managed by FIFA who could then assure that the exclusive rights for ground-advertisement space were safeguarded for FIFA sponsors. Lenskyj (2010, p. 19) makes a similar point in relation to the IOC, noting how they have created a 'moral barrier' whereby they use the arm's length principle when issues get difficult, willing to state that they have little control over a country's policies and practices, such as when issues of HR abuses arise: yet this is the same organisation, that can demand significant tax concessions and insistence that governments give commitments to comply with their rules policies, such as banning protests in or near Olympic venues, or strict controls of advertising space near venues.

8. CONCLUSION

What emerges is a complex and often contradictory picture of how of sport events can impact on peoples HR. What this discussion illustrates is that the impact of sport events on HR will always be mixed. The larger the event, the easier it is to find evidence to fit a particularly standpoint. The impression that one is left with is paradoxical; between the academic community, the sport practitioners, politicians, and governing bodies, where different sides can take a pick and mix approach to the selection of evidence. The practitioner can continue the constant refrain of sport as a means to express rights, or enhance and strengthen them – so that the benefits are freely transferred between sport activities in general, to sport events in particular. In contrast, the academic community can appear to be the voices in the wilderness, whereby the critical evidence that sport can have little or no impact, or can at times be detrimental to rights, is simply ignored. The weakness of the theoretical representation is also why there can be a pick and mix approach in terms of selecting examples to fit a particular position of sport events being a positive or negative force. Indeed, on this point we agree with Chalip (2006, p. 109) that the social value of sport events may well be a 'core source of event value' and with Preuss (2007) who argued that that it is difficult to identify event legacy as either having a positive or a negative value. Indeed we argue that HR as a key aspect of social value should not be left to serendipity, and part of the rationale for the model developed here is to provide a more nuanced appreciation of HR and sport events.

HR are rarely stated explicitly as tangible specific legacy outcomes of a sport event. According to Preuss (2007) legacy cube, outcomes tend to fall into the intangible, unplanned category. The reasons why they are important to consider is because not only are they implicit in many other legacy outcomes, they are also prone to greater scrutiny but paradoxically harder to define and measure. Sport organisations are increasingly engaging with the language of sport for good with its rhetorically thick language of values, such as respect, equality, tolerance, and understanding, whilst ignoring the contradictions. Whilst quick to use this rhetoric, sporty organisations can sometimes be just as quick to deny they have any power to bring about change when presented with some uncomfortable truths of HR abuses. It is a position which lacks credibility and, as Hoffman and McNulty (2009) argue, is often an excuse to justify bad business behaviour. There is also evidence to support Cmiel criticism of how sport organisations and politicians continue to be duplicitous in 'ritually mouthing slogans they ignore when convenient' (2004, p. 118).

The model developed in this paper provides a framework from which to review the potential impacts on HR from an events' perspective. A crucial point of the model is to illustrate how and why sport events create the *potential* to impact on HR: whether or not they do depends on how effectively key stakeholders use the legacy levers to prevent rights being pulled away, or to positively push them. The model makes a contribution to any exercise that seeks to make the measurement of rights more tangible, however there is much more to be done. Certainly in outlining the model, the next stage is to apply it to a specific event or events, with particular attention given to the sources of information that can give a clearer measurement of the outcomes and outputs. Used in this way the model can serve as a further buttress to subsequent discussions concerning sport events, HR, and the effective leveraging of sport events to improve people's rights. Indeed it is possible to concur with Coalter that at present the consensus is more that sport may have the capacity to elicit change, but that the evidence at the moment is still inconclusive at best (Coalter, 2010).

DISCLOSURE STATEMENT

No potential conflict of interest was reported by the authors.

REFERENCES

1 Adams, A., & Harris, K (2014). Making sense of the lack of evidence discourse, power and knowledge in the field of sport for development. International Journal of Public Sector Management, 27(2), 140–151.

2 Amnesty International. (2013). Qatar: End corporate exploitation of construction workers. Retrieved March 23, 2013 fromhttp://www.amnesty.org/en/news/qatar-end-corporate-exploitation-migrant-construction-workers-2013--11--17.

3 Angle, S. (2002). Human rights and Chinese thought: A cross cultural inquiry. Cambridge: Cambridge University Press.

4 Bairner, A., & Molnar, G. (2010). The politics of the Olympics: An introduction. In A. Bairner & Molnar, G. (Eds.), The politics of the Olympics. (pp. 3–14) London: Routledge.

5 BBC. (2014). Why London won the Olympics. Retrieved fromhttp://news.bbc.co.uk/sport1/hi/other_sports/olympics_2012/4618507.stm

6 Booth, K. (1999). Three tyrannies. In T. Dunne & N. J. Wheeler (Eds.), Human Rights in Global Politics (pp. 31–70). Cambridge: Cambridge University Press.

7 Booth, D., Kidd, B , & Toohey, K. (2001). Three responses to the IOC's award of the 2008 Olympics to Beijing. Australian Society of Sport Historians (ASSH) Bulletin no. 34, pp. 24–30.

8 Brackenridge, C. H. (2001). Spoilsports: Understanding and preventing sexual exploitation in sport. London: Routledge.

9 Brown, C. (1999). Universal human rights: A critique. In T. Dunne & N. Wheeler (Eds.), Human Rights in Global Politics (pp. 103–127). Cambridge: Cambridge University Press.

10 Cense, M., & Brackenridge, C. (2001). Temporal and developmental risk factors for sexual harassment and abuse in sport.European Physical Education Review, 7(1), 61–79.

11 Chalip, L. (2006). Towards social leverage of sport events. Journal of Sport and Tourism, 11(2), 109–127.

12 Cmiel, K. (2004). The recent history of human rights. The American Historical Review, 109(1), 117–135.

13 Coakley, J. (2011). What counts as "positive development?". Journal of Sport & Social Issues, 35(3), 306–324.

14 Coalter, F. (2007). A wider social role for sport: Who's keeping the score? London: Routledge.

15 Coalter, F. (2010). The politics of sport-for-development: Limited focus programmes and broad gauge problems?International Review for the Sociology of Sport, 45(3), pp. 295–314.

16 Coalter, F. (2013). Sport for development: What game are we playing. London: Routledge.

17 COHRE. (2008). One world, one dream: Housing rights violations and the Beijing Olympic Games. Geneva: COHRE. Retrieved from www.cohre.org

18 Collins, M. (2010). From ‘sport for good’ to ‘sport for sport's sake’ – not a good move for sports development in England. International Journal of Sport Policy, 2(3), 367–379.

19 Donnelly, P. (2008). Sport and human rights. Sport and Society, 11(4), 381–394.

20 Donnelly, P., & Petherick, L. (2006). Workers' playtime: Child labour at the extremes of the sporting spectrum. In D. McArdle & R. Giulianotti (Eds.), Sport, civil liberties and human rights (pp. 9–29). London: Routledge.

21 Dowling, M., Robinson, L., & Washington, M. (2013). Taking advantage of the London 2012 Olympic games: Corporate responsibility through sport partnerships. European Sport Management Quarterly, 13(3), 269–293.

22 Eick, V. (2010). A neoliberal sports event? FIFA from the Estadio Nacional to the fan mile. City: Analysis of Urban Trends, Culture, Theory, Policy, Action, 14(3), 278–297.

23 Fasting, K., Brackenridge, C., & Kjølberg, G. (2013). Using court reports to enhance knowledge of sexual abuse in sport. Scandinavian Sports Studies Forum, 4, 49–67.

24 Gervis, M., & Dunn, N. (2004). The emotional abuse of elite child athletes by their coaches. Child Abuse Review, 13, 215–223.

25 Giulianotti, R. (2004). Human rights, globalization and sentimental education: The case of sport. Sport in Society: Cultures, Commerce, Media, Politics, 7(3), 355–369.

26 Giulianotti, R. (2006). Human rights, globalisation and sentimental education. In R. Giulianotti & D. McArdle (Eds.), Sport, civil liberties and human rights (pp. 63–77). London: Routledge.

27 Gratton, C., & Preuss, H. (2008). Maximising Olympic impacts by building up legacies. The International Journal of the History of Sport, 25(14), 1922–1938.

28 Griffin, J. (2008). On human rights. Oxford: OUP.

29 Grix, J., & Carmichael, F. (2012). Why do governments invest in elite sport? A polemic. International Journal of sport policy and politics, 4(1), 73–90.

30 Hartmann, D., & Kwauk, C. (2011). Sport and development: An overview, critique and reconstruction. Journal of Sport and Social Issues, 35(3), 284–305.

31 Hartill, M. (2009). The sexual abuse of boys in organized male sports. Men and Masculinities, 12(2), 225–249.

32 Hoffman, W. M., & McNulty, R. E. (2009). International business, human rights, and moral complicity: A call for a declaration on the Universal rights and duties of business. Business and Society Review, 114, 541–570.

33 Horne. J. (2010). The politics of hosting the Olympic games. In A. Bairner & G. Moner (eds.), The Politics of the Olympics (pp. 27–40). London: Routledge.

34 Hough, P. (2008). Make goals not war: The contribution of international football to world peace. The International Journal of the History of Sport, 25(10), 1287–1307.

35 Hunt, L. (Ed.). (1996). The French revolution and human rights: A brief documentary history. Boston, MA: Bedford/St. Martin's.

36 IOC. (2014). Development through sport: Humanitarian assistance through sport, Retrieved May 12, 2014 fromhttp://www.olympic.org/news/humanitarian-assistance-through-sport/209698.

37 Jennings, W., & Lodge, M. (2009). Tools of security risk management for the London 2012 Olympic Games and FIFA 2006 World Cup in Germany (Discussion Paper No. 55). London: Centre for Analysis of Risk and Regulation, LSE

38 Kay, T. (2009). Developing through sport: Evidencing sport impacts on young people. Sport in Society: Cultures, Commerce, Media and Politics, 12(9), 1177–1191.

39 Kellett, P., Hedeb, A.-M., & Chalip, L. (2008). Social policy for sport events: Leveraging (relationships with) teams from other nations for community benefit. European Sport Management Quarterly, 8(2), 101–121.

40 Kelly, S., & Waddington, I. (2006). Abuse, intimidation and violence as aspects of managerial control in professional soccer in Britain and Ireland. International Review for the Sociology of Sport, 41(2), 147–164.

41 Kerr, G. (2010). Physical and emotional abuse of elite child athletes: The case of forced physical exertion. In C. H. Brackenridge & D. Rhind (eds.) Elite child athlete Welfare: International perspectives (pp. 41–50). West London: Brunel University Press.

42 Kerr, G .A., & Stirling, A. (2008). Child protection in sport: Implications of an athlete-centred philosophy. Quest, 60, 307–323.
43. Kerr, G. A., & Stirling, A. E. (2012). Parents' reflections on their child's experiences of emotionally abusive coaching practices. Journal of Applied Sport Psychology, 24(2), 191–206.

43 Kidd, B., & Donnelly, P. (2000). Human rights in Sports. International Review for the Sociology of Sport, 35(2), 131–148.

44 Kissoudi, P. (2008). The Athens Olympics: Optimistic legacies – Post-Olympic assets and the struggle for their realization.The International Journal of the History of Sport, 25(14), 1972–1990.

45 Lang, M. (2010). Intensive training in youth sport: A new abuse of power? In K. P. Vanhoutte & M. Lang (eds.), Bullying and the abuse of power: From the playground to international relations (pp. 57–64). Freeland: Inter-Disciplinary Press.

46 Lauren, P. G. (2011). The evolution of international human rights: Visions seen (3rd ed.). Philadelphia: University of Pennsylvania Press.

47 Lenskyj, H. J. (2006). The Olympic industry and civil liberties: The threat to free speech and freedom of assembly. In R. Giulianotti & D. McArdle (Ed.), Sport, civil liberties and human rights (pp. 78–92). London: Routledge.

48 Laureus Sport for good Foundation, Foundation overview (2013). London: Laureus Sport for good Foundation.http://www.laureus.com/sites/default/files/publications/foundation_book_november_2013_web_spreads.pdf.

49 Lenskyj, H. (2010). Olympic power, Olympic politics: Behind the scenes. In A. Bairner & G. Moner (Eds.), The Politics of the Olympics (pp. 15–26). London: Routledge.

50 Levermore, R. (2011). Evaluating sport-for-development. Progress in Development Studies, 11(4), 339–353.

51 Preuss, H. (2007). The conceptualisation and measurement of mega sport event legacies. Journal of Sport & Tourism, 12(3–4), 207–228.

52 Rowe, N. (2005). Keynote paper: How many people participate in sport? The politics, practice and realities of measurement – the

English experience. International association for sports information. Beijing: Beijing Sport University.

53 Schneider, A. (2004). Privacy, confidentiality and human rights in sport. Sport in Society: Cultures, Commerce, Media, Politics,7(3), 438–456.

54 Schulenkorf, N. (2010). Sport events and ethnic reconciliation: Attempting to create social change between Sinhalese, Tamil and Muslim sportspeople in war-torn Sri Lanka. International Review for the Sociology of Sport, 45(3), 273–294.

55 Sen. (2002). Rationality and freedom. Cambridge, MA: Harvard University Press.

56 Sen, A. (2004). Elements of a theory of human rights. Philosophy and Public Affairs, 32(4), 315–356.

57 Smith, D. (2010, April 1). Life in 'Tin Can Town' for the South Africans evicted ahead of world cup. The Guardian.http://www.guardian.co.uk/world/gallery/2010/apr/01/world-cup-2010-south-africa?picture=361067380

58 Smith, A., & Fox, T. (2007). From 'event-led' to 'event-themed' regeneration: The 2002 commonwealth games legacy programme. Urban Studies, 44(5–6), 1125–1143.

59 Spaaij, R. (2009). The social impact of sport: Diversities, complexities and contexts. Sport in Society, 12(9), 1109–1117.

60 Stirling, A. E., & Kerr, G. A. (2007). Elite female swimmers' experiences of emotional abuse across time. Journal of Emotional Abuse, 7(4), 89–113.

61 Stirling, A. E., & Kerr, G. A. (2009). Abused athletes' perceptions of the coach athlete relationship. Sport in Society, 12(2), 227–239.

62 Sugden, J. (2010). Critical left-realism and sport interventions in divided societies. International Review for the Sociology of Sport, 45(3), 258–272.

63 Timms, J. (2012). The Olympics as a platform for protest: A case study of the London 2012 'ethical' Games and the Play Fair campaign for workers' rights. London: LSE.

64 United Nations. (2014). Sport for development: Why sport. Retrieved fromhttps://www.un.org/wcm/content/site/sport/home/unplayers/fundsprogrammesagencies

65 Valentini, L. (2012). What sense are human rights political? A preliminary exploration. Political Studies, 60(1), pp. 180–194.

CHAPTER 5

An Exploration of the Relationship between Large-Scale Sporting Events and Education for Sustainable Development: The Case of the Melbourne 2006 Commonwealth Games

Rob Harris[a]

[a] Australian Centre for Event Management, University of Technology, Sydney, PO Box 222, Lindfield, 2070, Sydney, Australia

ABSTRACT

This study explores the relationship between large-scale sporting events (LSSEs) and education for sustainable development (ESD). A case study-based explorative research approach was employed utilising the 2006 Melbourne Commonwealth Games (MCG) – one of only two LSSEs that have taken place in Australia (the other being the Sydney 2000 Olympic Games) which have purposely sought to engage with a sustainable development agenda. The enquiry concludes that the MCG acted as a vehicle for ESD and in so doing impacted, to varying degrees, organisations, groups and individuals across its host community (the State of Victoria). Various factors were identified that impacted the ESD process in this context. Facilitating factors included the presence of an environmental strategy inclusive of ESD elements; a pre-existing commitment by government to sustainable development; and the use of partnerships by the event's organisers to progress its environmental agenda. Constraints on the ESD process were also identified. Central amongst these were the failure to embrace environmental considerations in the MCG's enabling legislation and the limited resources given over to, and lack of emphasis placed upon, the event's overall environmental programme.

KEYWORDS

sustainable development, education, large-scale sporting event, environment, community, Commonwealth Games

1. INTRODUCTION

The concept of sustainable development emerged in the 1970s and 1980s. Its origins lay in concerns within industrialised nations that patterns of production and consumption were not sustainable in terms of the earth's capacity to support them. Today, the challenge embodied in the concept of sustainable development represents, as Kofi Annan, former Secretary General of the United Nations (UN), has noted, 'our biggest challenge in this new century ...'.1 Education plays a central role in humanity's response to this challenge, as is evidenced by the UN General Assembly's declaration of the *Decade of Education for Sustainable Development* (2005–2014). In the

Australian context, this role has also been acknowledged in various planning documents that have been developed at the federal level2 and the state level.3 These documents explicitly recognise that education for sustainable development (ESD) can occur in a multitude of non-institutional settings. It is the intent of this exploratory study to examine one such setting, large-scale sporting events (LSSEs).

Over the past decade, there has been an emerging acceptance by both sporting bodies with ownership of LSSEs and the communities that host them that these activities have the capacity to be leveraged for ESD purposes. This acknowledgement, however, has not been accompanied by any discernible interest by researchers. This paper seeks to go some way towards redressing this situation by exploring the process of host community ESD through examining a specific LSSE – the 2006 Melbourne Commonwealth Games (MCG). This event was chosen as the focus of this paper as it is the most recent of only two LSSEs (the other being the 2000 Sydney Olympic Games, SOG) that have taken place in the Australian context that have made a significant effort to engage with a sustainable development agenda, inclusive of ESD elements.4 This being the case, the MCG presents an opportunity to learn about a matter where theory and research are lacking.

2. STUDY BACKGROUND

2.1 Education for Sustainable Development

The concept of sustainable development gained increased currency internationally after the UN Earth Summit in 1992. This event saw 178 nations commit to a sustainable-development action plan (Agenda 21). This plan explicitly acknowledged the importance of education to efforts in this area noting that it was:

... indispensable to changing people's attitudes so that they have the capacity to assess and address their sustainable development concerns. It is also critical for achieving environmental and ethical awareness, values and attitudes, skills and behaviour consistent with sustainable development and for effective public participation in decision making.5

While such statements make clear the importance of education in the process of sustainable development, the concept of ESD itself remains ill-defined. A review of the many definitions proposed by various organisations and researchers6nonetheless indicates that there has been a change over time in how the concept has come to be understood. Initially, ESD was concerned simply with knowledge, attitudes and values as regards the environment. More recently, this focus has been expanded to include the engagement of individuals and organisations in a dialogue around social change for sustainability. Indicative of such latter perspectives is that of Fein:

Education for sustainability integrates ecological thinking and the wise use of natural resources – conservation – with the equally important concerns of social, economic and political sustainability. The aim of education for sustainability is to develop skills that can enable all citizens and, through them, our social institutions, to play a role in the transition to sustainability. As such, it encompasses a vision for society that is not only ecologically sustainable but also socially, economically, and politically sustainable.7

Viewed in this way, ESD can be seen as a multifaceted educational process that seeks to develop a community's capacity to link economic, political and social processes to the imperative of ecological sustainability. At its base, the process still requires awareness, understanding and a capacity to engage with sustainable practices, as well as attitudes and behaviour conducive to such engagement, but these factors, while necessary, are not sufficient for ESD to occur. ESD also requires citizens to have the skills and knowledge necessary to challenge and reshape their societies in order to make them more sustainable not only environmentally but also economically and socially. The degree to which host communities can potentially leverage LSSEs as a mechanism by which they can achieve these outcomes is a matter that this study seeks to provide insights into.

2.2 Sport and Sustainable Development and Education for Sustainable Development

The link between sport and sustainable development has only recently begun to be explored by researchers, with Lenskyj noting that prior to 1998 this association had received little attention in the academic literature.8 This situation appears not to have changed significantly over the following decade with Mallens, Stevens and Adams concluding that the relationship between sport and the environment remains a little explored area.9

While researchers have shown only a limited interest in the relationship between sport and the environment, various organisations, event owners and host communities have been active in developing this linkage. The United Nations Environment Programme (UNEP), for example, has produced a long-term strategy on sport and the environment (*Michezo*).10This strategy seeks to promote sustainable development through sport and in so doing leverage the 'popularity of sports to promote environmental awareness and respect for the environment amongst the public, especially young people'.11 The Council of Europe's *Code of Sustainability in Sport* also recognises the power of sport in this area and calls for, amongst other things, the development of curricula and the creation of information packs for environmental education.12 Green Cross Spain (an environmental non-government organisation, NGO) provides a further example of a body that has acknowledged sport's potential to be an agent for ESD. This body joined with the Spanish Ministry for the Environment, the Spanish Olympic Committee and the Spanish Biodiversity Foundation to produce the *Spanish National Strategy on Sport and Sustainability*. This strategy sought to develop an understanding of sustainable development within sporting bodies, as well as enhance the capacity of sport to raise environmental awareness in the broader community.13 Other NGOs are also observed to have been active in seeking to further develop the relationship between sport and ESD. The Global Sports Alliance, for example, is dedicated to promoting environmental awareness and action through sport.14 The environmental NGOs Greenpeace and the World Wide Fund for Nature (WWF) have also played an active role in providing input into and/or making assessments of the environmental component of summer and winter Olympic Games, including elements that can be linked to ESD.15

Through their policies, declarations and practices, it is evident that some sporting organisations appreciate the role sport can play in broader societal efforts directed at sustainable development. Of particular note in this regard has been the progressive engagement of the International Olympic Committee (IOC) with the concept of sustainability. This engagement can arguably be traced back to its signing of the Earth Pledge in 1992, a document which emerged out of the Earth Summit in Rio de Janeiro that same year.16 The involvement of the IOC in this area has become deeper over time, as is reflected in its decision to create a Sport and Environment Commission, which, since its establishment in 1995, has conducted nine conferences exploring the link between the environment and sport, with the 10th scheduled for October 2013 in Sochi, Russia.17 The IOC has also acted to change its charter, making the environment the third pillar of Olympism (the others being sport and culture).18

The IOC has not been alone in recognising the potential benefits LSSEs offer for ESD. Another major sports body that has been active in this area is the Fédération Internationale de Football Association (FIFA). Beginning in 2006, FIFA introduced an environmentally focused programme of actions entitled *Green Goal,* which they developed in concert with the UNEP. The latest iteration of this programme (in association with the 2010 FIFA World Cup in South Africa) included a component on environmental education and public awareness.19

It is also observed that LSSE host communities do not necessarily need to be 'pushed' by event owners in order to leverage such events for sustainable-development purposes. For example, the Government of Macao when hosting the 4th East Asian Games in 2006 faced no requirement by the event's owner, the East Asian Games Association, to engage in an environmental programme, yet they nonetheless opted to do so. Significantly, from an ESD perspective, an aspect of their programme involved showcasing environmental best practice and undertaking activities intended to raise environmental awareness in order to encourage a 'greener' and healthier lifestyle amongst its citizens.20 Other examples of host governments acting in this way can also be identified, including the State Government of New South Wales which included educational elements within its overall environmental programme for the Sydney 2000 Olympic Games.21 The degree to which the LSSE host community that is of interest in this study (the State of Victoria, Australia) saw in the MCG an opportunity to progress its ESD agenda is a matter that this study will seek to shed light upon.

2.3 The Commonwealth Games and ESD

The Commonwealth Games (CG) is a multi-day sporting event open to nations who are members of the Commonwealth.22Ultimate responsibility for the conduct of this event rests with the Commonwealth Games Federation (CGF).23 In order for a city to host the CG, it must engage in a competitive bidding process based upon a defined set of criteria provided by the CGF. Significantly, from the perspective of this study, these criteria necessitated little by way of response from cities bidding for the 2006 CG in the area of environmental sustainability.24 This situation has changed in more recent times with the last revision of the CG's bid document (May 2011) requiring significantly more detail in this area, including information on how the event will be used for raising environmental awareness.25

CG host nations, prior to the conduct of the MCG, also appear to have made little effort to independently pursue an environmental agenda through their delivery of the event. The final report of the Manchester 2002 Commonwealth Games, for example, refers to the environment only in the context of waste management and limits discussion of environmental education matters to this narrow context.26 Post-MCG, however, this situation has changed, with host cities becoming more proactive in this area. The organisers of the Delhi Commonwealth Games worked with the UNEP to create a number of environmental codes,27 and the successful 2014 host city, Glasgow, developed a detailed sustainability plan that it launched in 2010.28 ESD practices have resulted from, or have been incorporated in, these initiatives.

The above discussion suggests that the decision by the Victorian State Government to leverage its hosting of the CG for sustainable development purposes was more likely linked to its own priorities than those of the CGF. It also highlights that the CGF, as well as past host cities, were unlikely to be sources of direction or guidance as regards its efforts in this area. Additionally, the actions of subsequent CG cities in the sustainability area indicate that the MCG could be viewed as a watershed event in terms of the degree to which sustainable development has come to feature in the planning and delivery of the Games.

2.4 Study Goals and Research Approach

By examining the MCG case, this study aims to provide insights into: the types of organisations that have the capacity to play a role in the leveraging of LSSEs for ESD purposes; programmes and other initiatives with the potential to generate host community ESD outcomes; and factors that may serve to facilitate or hinder efforts to progress an ESD agenda via an LSSE.

Given research and associated theory were found to be lacking in the context of the matters that this study addresses, an exploratory research approach was employed utilising a single case study with the capacity to provide meaningful insights into the issue at hand. Case studies used in this way can be viewed as 'instrumental' in nature in that the case itself is of secondary interest, with their examination serving to advance understanding of a broader issue – in this case ESD in LSSE settings.29 Further, used in this way, case studies have the potential to act as disseminators of innovative practice.30

In seeking to learn from the MCG case, multiple sources of information were engaged with including documentary, audio-visual and archival records. Information gained from these sources both provided input into the study itself and served as a means of stimulating paths of inquiry that were later explored through personal interviews.

Due to the previously noted lack of prior research, and published material more generally, concerning the link between LSSEs and ESD, face-to-face interviews with key informants contributed significantly to this enquiry. Data gathered in this way, as Gillham notes, have the capacity to expand upon documentary evidence, gauge its 'truth' and provide a critical perspective of it.31 Additionally, interviews were used to identify sources of information beyond those that had previously been drawn upon.

The approach used to select interviewees can be classified as what Neuman calls purposeful, strategic or judgement-based with its intent being to identify individuals with the best capacity to inform the study.32 In the first instance, interviewees were identified using the data sources noted previously. Later, informants themselves were asked to suggest other individuals that they believed could inform the research, a technique Patton describes as 'snowballing'.33 While this technique often resulted in those already identified being named, it nonetheless did serve to expand the pool of interviewees beyond those generated from secondary sources. In total 20 people were interviewed, all of whom opted not to remain anonymous. The list of interviewees and their titles/positions at the time of the MCG is provided in Table 1.

Host-community ESD through the medium of LSSEs has been conceived of in this study as a multi-stakeholder process, with those organisations or groups actively involved in the process being categorised as either focal, collaborative or independent. Focal organisations are defined here as those bodies that served as the primary drivers of the ESD process, while collaborative stakeholders are those organisations with whom they sought to cooperate in order to generate ESD-related outcomes. The remaining stakeholder type – independent stakeholders – are those organisations or groups whose contributions to the ESD process was direct, as opposed to through the medium of a cooperative relationship with a focal organisation.

Regarding the applicability of the study's findings to other contexts, it needs to be acknowledged that there are difficulties in generalising from case studies due to their lack of representativeness. 34

However, a number of researchers35 argue that it is possible to draw lessons/perspectives from cases that can refine action in similar circumstances and/or result in the capacity to make 'modest speculations' as to the likely applicability of their findings to other similar cases. It is this later view that is taken in this study.

Table 1 Melbourne 2006 Commonwealth Games interview list.

Name	Position at time of Melbourne Commonwealth Games[a]
Cheryl Batagol	Member, Environment Committee, Melbourne Commonwealth Games and Chair, Melbourne Water
Sheridan Blunt	Savings in the City: Green Hotel Program Project Co-ordinator, City of Melbourne
Dr Michael Buxton	Member, Environment Committee, Melbourne Commonwealth Games and Associate Professor Environment and Planning, RMIT
Lucy Callaghan	Programme Manager for Environment, Office of Commonwealth Games Co-ordination
Karen Cosson	Litter Champion, Victorian Litter Action Alliance
Christine Forster AM	Member (Chair), Environment Committee, Melbourne Common wealth Games, and Deputy Chair, Victorian Water Trust Advisory Council
Rob Gell	Environment Ambassador, Commonwealth Games and Executive Chairman of Access Environmental Pty Ltd
Marcus Godhino	Member, Environment Committee, Melbourne Commonwealth Games and Executive Director, Environment Victoria
Debra Gonsalvez	Manager, Renewable Energy, Department of Sustainability, State Government of Victoria
James Holgate	Manager, Communications, Toyota, Port Melbourne
Naomi Lind	Programme Development Officer, Commonwealth Games, Department of Education
Gillian McDonald	Co-ordinator, Our River Our Games Program, Melbourne Water
Kate Mckeand	Project Officer, Environment Programme, Office of Common-wealth Games Co-ordination
David Meiklejohn	Manager TravelSmart, Department of Infrastructure, Victoria
Patrick Ness	Architect and Design Director, Cox Architects
Kendrea Pope	Manager, Education Programme, Office of Commonwealth Games Co-ordination
Rose Read	Team Leader Community Programmes, Sustainability Victoria
Jenny Sams	Director, Office of Commonwealth Games Co-ordination
Adrienne Stephens	Executive Officer, ECO-Buy, Victoria
Hugh Wareham	Director, Office of Victorian Commissioner for Environmental Sustainability

[a] Positions given in this table are those interviewees held at the time of the Melbourne 2006 Commonwealth Games.

3. RESULTS

The narrative that emerged from employing the previously cited research approach was extensive. Discussion here centres on its key findings and the potential insights such findings provide as to the functioning of the ESD process in LSSE settings.

3.1 Active Stakeholders in the ESD Process

3.1.1 Focal Stakeholders and Their Collaborators

The key focal organisation in the ESD process was identified as the Office of Commonwealth Games Coordination (OCGC).36This body's actions in the environmental area were guided by the Victorian Government's desire to use the Melbourne 2006 Commonwealth Games as a means of '... building on Victoria's clean and green image and reputation for liveability, innovation and leadership in environmental policy and leaving a legacy of environmentally sustainable assets and behaviour'.37 In pursuit of these goals, the OCGC developed a formal environmental sustainability strategy, which aimed to ensure the event was delivered in a way that was 'wastewise, carbon neutral and waterwise'.38 Included in this strategy were a range of objectives linked to ESD. These objectives encompassed raising environmental awareness, environment-related education/training and attitude and behaviour change regarding environmental matters (e.g. litter, public transport usage).39To achieve these outcomes, the OCGC acted either directly or by establishing collaborative stakeholder relationships with organisations in the government, non-government organisation and community sectors (see Table 2 for a listing of these organisations).

Table 2 Office of Commonwealth Games Co-ordination ESD collaborating stakeholders.

Government organisations	
Sustainability Victoria	A statutory authority that acts to support the delivery of the Victorian Government's Framework for Environmental Sustainability. In performing this role, Sustainability Victoria provides advice and assistance to the government on sustainable resource use with a view to minimising greenhouse gas emissions and resource usage by business and individuals.[a]
Environment Protection Agency (EPA)	A Victorian Government statutory body that acts to address issues associated with land, air, water and noise pollution. Additionally, the EPA is concerned with developing sustainable solutions to waste management. In performing these functions, the EPA acts to form partnerships with business, the community and government.[b]
Melbourne Water	A Victorian Government instrument which is operated by an independent board of directors. Its primary functions are to manage Melbourne's water supply catchments, remove and treat sewage and manage rivers, creeks and major drainage systems within its geographic area of responsibility.[c]
Parks Victoria	A statutory authority with responsibility for the management of the state's parks, reserves and other land under state control.[d] Parks Victoria also has responsibility for the recreational management of a number of waterways in close proximity to Melbourne, specifically Port Phillip Bay, Westernport and the Yarra and Maribyrnong rivers.[e]
The Department of Sustainability and Environment	The state of Victoria's main government agency responsible for the sustainability of both the natural and built environment.[f] Its brief is wide ranging and encompasses water management; stewardship of public land; forest fire management; encouragement of sustainable resource use and management practices by business and the broader community; climate change and greenhouse policy and sustainability policy development.[g]
Major Projects Victoria	A State Government agency that manages large-scale development and construction projects engaged in by the Victorian Government.[h]
Catchment Management Authorities	Statutory bodies, of which there are 10 in Victoria, whose role is to ensure integrated land and water management within their designated area. To achieve this goal, they work in association with local communities, government, landholders and industry across the catchments for which they are responsible.[i]
ECO-Buy Limited	Established in 2002, ECO-Buy Limited is a not-for-profit company funded by the Victorian Department of Sustainability and Environment. Its function is to encourage the purchase of green products by business and government.[j]
The Department of Education and Training (now the Department of Education and Early Childhood Development)	A government department responsible for the development of policy on school education and the delivery of primary and secondary education in Victorian Government schools.[k]

City of Melbourne Council (CMC)	CMC is a public statutory body established to govern the Municipality of Melbourne.[l]
Non-government organisations	
Greening Australia	A not-for-profit non-government organisation that operates on an Australia-wide basis. Greening Australia's primary role lies in vegetation protection and restoration, a role that it pursues via the creation of partnerships with local communities, land-holders, government and business.[m]
Conservation Volunteers	A not-for-profit non-government organisation that acts to attract, place and manage volunteers in conservation projects Australia-wide.[n]
Clean Up Australia	A not-for-profit organisation and non-government organisation whose original focus was litter and waste and the engagement of communities around these environmental concerns. In more recent years, this focus has expanded to embrace water and climate change, and along with this expansion has come a widening of its geographic scope of operation with the creation in 1992 of Clean Up the World.[o]
The Victorian Litter Action Alliance (VLAA)	The peak body for litter management and reduction in Victoria. This body draws its membership from state and local government, industry associations and the community. VLAA's primary function is to ensure a coordinated approach to litter prevention in Victoria.[p]
Waterwatch Victoria	A component of a national volunteer-based community water-quality monitoring network. Waterwatch's functions extend to providing water quality and biological monitoring training and support to volunteers, schools and community groups; supporting education in the area of water monitoring in schools and the broader community; engaging in awareness-raising activities; and developing partnerships with government, community groups and private firms with the intent of improving river health.[q]

[a] Sustainability Victoria, *Our Business*.
[b] EPA, "About EPA Victoria."
[c] Melbourne Water, *Melbourne Water*.
[d] Parks Victoria, *Annual Report 2006–07*.
[e] OCGC, *Melbourne 2006 XVIII Commonwealth Games*, 9.
[f] DSE, "Sustainable Household Challenge."
[g] Ibid.
[h] Major Projects Victoria, "About Major Projects Victoria."
[i] See note 42 above.
[j] See note 82 above.
[k] DEECD, "About Us."
[l] CMC, *City of Melbourne Annual Report*, 19.
[m] Greening Australia and Green Corps, *Green Corps and the Environment*.
[n] Conservation Volunteers, "About Us."
[o] Clean Up the World, "The Clean Up Story."
[p] VLAA, "Home."
[q] Waterwatch Australia, "About Waterwatch."

In addition to the OCGC, the Melbourne 2006 Commonwealth Games Corporation (M2006) performed a 'weak' focal role with its actions being limited to joint development with the OCGC of the event's environmental procurement guidelines, waste-management practices and workforce training which was inclusive of environmental elements.40 In terms of M2006's formal environmental responsibilities, its establishment act simply notes that it was responsible for delivering ' ... environmental benefits to Victorians and Australians'.41

3.2 Independent Contributors

Independent contributors were identified as the Victorian Government's Department of Infrastructure; Melbourne City Council (MCC); the media; the Society for Sustainability and Environmental Engineering; Environment Institute of Australia and New Zealand; and Waste and Management Association of Australia.

3.3 ESD Programmes/Initiatives

3.3.1 Office of Commonwealth Games Coordination

Environmental procurement policy. In October 2006, an environmental procurement policy was developed by the OCGC in concert with M2006.The objectives of this policy were to promote 'environmental awareness, positive environmental performance and continuous environmental improvement to potential Commonwealth Games suppliers'.42 This policy emerged from the work of a task force (Environmental Procurement Taskforce) created specifically for this purpose. Businesses submitting a tender or proposal for the supply of nearly all services and products to the MCG were required to engage directly with this policy. This requirement in turn necessitated that they show compliance with all relevant environmental legislation and regulations, along with details of their environmental policies, plans and how these were being implemented in order to engender positive environmental outcomes. When assessing compliance with this policy, evidence was sought, as appropriate, regarding practices in a range of areas, including conservation of resources; continual improvement through the establishment, meeting and exceeding of environmental objectives; reduction and management of environmental impacts resulting from operational practices and/or products; and involvement/education of staff, clients, suppliers and/or the general community regarding environmental matters.43

From an ESD perspective, the need to provide this evidence would arguably have required a significant organisational learning effort for many businesses.

Training programmes. The OCGC and M2006 provided all Games staff and volunteers (15,000) with the publication *Team2006 Workbook*, along with an abridged version of this publication (*The Pocket Guide*).44 Both of these publications dealt with approaches employed by the OCGC to deal with the three major environmental objectives of the event, specifically being wastewise, carbon neutral and waterwise. Training and induction sessions drew heavily upon the *Team2006 Workbook*, thus ensuring the event's environmental programme featured prominently in these sessions. Environmental messages were further reinforced by a workforce newsletter entitled *United*. In the case of staff and volunteers assigned to venues, specific training was provided that dealt with venue environmental policies and practices, such as those associated with waste management, packaging and recycling.45

Production of ESD-linked resources and communication practices. In the process of pursuing an environmental programme through the MCG, the OCGC generated a number of publicly available documents and electronic resources concerning aspects of its environmental programme. Additionally, the OCGC undertook other practices, specifically site tours, presentations and media liaison activities, which also served to communicate the nature of its environmental programme to various audiences. These efforts, from an ESD perspective, can be viewed, according to Dr Michael Buxton, as playing a role in raising awareness and/or increasing the knowledge base of specific groups/individuals for whom they were intended, along with the community more broadly.46

Collaborative programmes and initiatives. The OCGC, as noted previously, sought to develop stakeholder relationships with a number of organisations in order to pursue its environmental agenda. A number of these collaborations resulted in programmes or initiatives that can be seen as linked to the process of host-community ESD via the MCG. These are discussed below.

Environmental Stakeholder Forum (ESF). Some three years prior to the start of the MCG, based on a recommendation made by the Commonwealth Games Advisory Committee – Environment (CGAC-E),47 the OCGC established an ESF. This body comprised 18 representatives from various government departments and agencies involved with the environmental aspects of the Games, as well as environmental groups and event sponsors.48 According to Rob Gell, when viewed purely from an ESD perspective, the Forum served to facilitate learning between the organisations involved regarding how the MCG and future LSSEs might be effectively leveraged for environmental purposes, including environmental education.49

The Commonwealth Games Education Program (CGEP). A joint initiative of the OCGC, Department of Education and Training, and Sustainability Victoria, the CGEP was guided by eight objectives, one of which concerned the uptake by school-aged children of environmentally sustainable behaviours. As a result of this objective, environment-related curriculum materials and supporting resources were developed and incorporated into the five components of the programme: 'Starting Blocks', 'Post Compulsory Resources', 'Let the Games Begin', 'United through the Games' and 'Compass'.50 To ensure widespread access to the CGEP, material associated with it was sent to all (approximately) 2400 schools in Victoria. In total, some 11,000 'Compass' teacher kits, 38,000 CD ROMs and 100,000 posters were distributed. Additionally, the programme was made available online through the M2006 website.51 Other actions that were taken in support of the education programme were the inclusion of an environmental supplement and poster in the main local broadsheet newspaper, *The Age*, and the creation of an online magazine for students – *CG Reps Gamezine*. It is noteworthy, from an ESD perspective, that the third edition of this e-zine (October 2005) was entitled '*Environment and You*' and contained an interview with the Melbourne 2006 Commonwealth Games Environment Ambassador (Rob Gell); a case study of a specific school's efforts to reduce, reuse and recycle; an overview of activities held for schools in association with the 'Our River Our Games' programme; a competition based around the event's mascot – the endangered South-eastern Red-tailed Black Cockatoo; and a list of environment-based activities and resources.52

As a way of assisting teachers in their efforts to engage with the event's education programme, a statewide professional development programme was conducted in all regions of Victoria.53 From an ESD standpoint, it is noteworthy that some of these professional-development sessions were conducted by the Victorian Association of Environmental Education. The sessions conducted by this body dealt with the conservation of the species represented by the event's mascot (a species of black cockatoo) and its stringy bark forest habitat.54

It is also of note that outside of the formal Games education programme some schools engaged their students in environmentally based volunteering activities linked to the event, including tree planting and litter collection.55 Additionally, according to Kate McKeand, a number of schools also participated in the CG *WasteWise Events Challenge*. This activity required schools to show how they had acted to minimise waste associated with an event that they had designed and delivered.56 To progress this initiative, the OCGC collaborated with Sustainability Victoria, linking it with their already established *WasteWise*School Program.57

Environmental mascot. The endangered South-eastern Red-tailed Black Cockatoo, as noted previously, was made the official mascot for the MCG and given the name Karak in recognition of its species' distinctive call. The Games was then leveraged to educate and focus the broader community's attention upon the importance of preserving this species' habitat. Actions taken with this intent in mind included the inclusion of educational material within the CGEP, promotional events conducted by the OCGC itself and the issuing of Karak number plates in association with Planet Ark and the Victorian Department of Sustainability and the Environment.58

Green Trades Program. The Green Trades Program was developed to provide training to plumbing and electrical apprentices and qualified trades people working on the construction of MCG venues and the Athletes Village. The focus of this programme was sustainable technologies and practices. Although it was a joint initiative between the OCGC and Sustainability Victoria, the cost of these training programmes was met by the OCGC. In total, 16 plumbing apprentices, 40 registered plumbers and 15 electrical apprentices took part in the programme.59

Waste-minimisation efforts. The OCGC and M2006, in association with Sustainability Victoria, sought to ensure the CG would be a low-waste event. Various strategies were employed for this purpose, including developing and implementing public and back-of-house recycling procedures; instituting packaging guidelines for merchandisers, suppliers, sponsors and other providers of goods and services; and conducting 'WasteWise' catering training for catering company managers and supervisors.60 Regarding this last point, resource material was also developed to aid catering companies in training their own staff in 'WasteWise' practices.61

Waste-minimisation efforts extended to construction waste resulting from the development and demolition of the major Games venues, specifically the Athletes' Village, Melbourne Cricket Ground and Melbourne Sports and Aquatic Centre. In connection with these venues, OCGC and Sustainability Victoria established a recycling benchmark of 95%, and worked with Sustainability Victoria and construction companies involved in these projects to establish systems to achieve this goal.62 From an ESD viewpoint, these actions can reasonably be argued to have acted to raise awareness, understanding and capabilities of organisations in the construction and catering industries in relation to the management of their waste streams.

Venue and Games Village design and construction initiatives. The OCGC, in association with Major Projects Victoria, required architectural firms and developers involved in the planning and construction of major Games infrastructure to incorporate water- and energy-saving technologies and design elements.63 To ensure this, sustainability covenants were developed and instituted in partnership with construction firms such as the Village Park Consortium, which was responsible for building the Games Village.64 In the context of water, and depending on the specific development, these design elements and technologies related to rainwater collection; water treatment and recycling; grey-water collection; landscaping; and the use of water-efficient appliances, fixtures and fittings.65

In order to achieve high levels of energy efficiency, permanent residential buildings at the Athletes Village were designed to a six-star standard, employing such design elements as insulation, air-leakage control, roof ventilation, energy-efficient lighting (including motion sensors), solar gas-boosted hot water systems and double glazing. Other major developments, specifically the Melbourne Cricket Ground modifications and the building of the Melbourne Sport and Aquatic Centre, also reflected the OCGC's commitment in this area. In the case of the former, 200 sqm of grid-connected photovoltaic panels, along with energy-efficient lighting systems, were installed, while the latter's construction included an 800-kW cogeneration power plant that used both gas and electricity.66

Viewed through the lens of ESD, these energy- and water-saving initiatives had the capacity to serve as a showcase for how architects and the construction industry as a whole might employ such technologies in future developments.67 Additionally, from the perspective of the general community, their presence in such high-profile Games venues may have acted to further reinforce messages associated with water and energy usage that they were receiving from other sources, such as power companies (see later discussion in this section). Athletes and team officials (of which there were 5867) were also likely to be impacted by such developments given the visibility of these initiatives within the Athletes Village, and the inclusion of information concerning them in the *Athletes Village Guidebook* that was provided to each athlete and team official.68

Our River Our Games Program and Yarra River litter barges. Many of the venues for the CG, such as the Melbourne Cricket Ground and the Melbourne Exhibition Centre, along with a number of paths and roads designated for use in athletic and bicycling events, lay within close proximity to the Yarra River. Additionally, the river was integral to the event's entertainment programme, serving as a 'stage' for both a nightly sound and light show and the opening and closing ceremonies.69 Given this, an opportunity was identified by the OCGC to leverage the attention the river would receive during the Games both to improve its health and, significantly from an ESD viewpoint, to increase community awareness of environmental issues associated with it.70 The programme the OCGC developed for these purposes was entitled 'Our River Our Games', and involved tree planting, night fauna studies, storm water education, litter clean-ups and drain stencilling activities.71Partnerships were crucial in the delivery of this programme, with a number of organisations involved in various aspects of it. Specifically, these organisations were Melbourne Water; Conservation Volunteers Australia; Department of Sustainability and Environment; Friends Groups; local councils; local schools; Parks Victoria; and Waterwatch Victoria. In total, the 'Our River Our Games' programme resulted in 41 individual activity days across 17 locations, with participation from some 200 adults and 230 school students.72 Implementing this programme was made easier, according to Gillian McDonald,73 because its main delivery partner, Melbourne Water, already had in place a strong community education programme based around the Yarra River and other Melbourne water courses.

While not a component of the 'Our River Our Games' programme, the purchase of two litter barges by the OCGC and Parks Victoria to ply the Yarra and Maribyrnong rivers, collecting rubbish from the two waterways, also served to create community awareness of the issue of litter as the Games approached. Painted prominently on these vessels were the words 'Commonwealth Games Environment Program' and 'Cleaning Our Waterways for the 2006 Commonwealth Games'.74

Commonwealth Games Anti-Litter Campaign. The Commonwealth Games Anti-Litter Campaign was a joint initiative of the OCGC, Sustainability Victoria and the Victorian Litter Action Alliance (VLAA). According to Karen Cosson, this campaign was based on extensive research into littering behaviour and attitudes towards littering.75 Significantly, from an ESD perspective, the focus of this programme was achieving behaviour change, with the MCG regarded as an opportunity to further raise community awareness of the littering issue and to 'achieve a legacy of ongoing litter-free behaviour'.76 This focus was acknowledged by Cheryl Batagol who noted that 'we were concerned much more with embedding behavioural change as opposed to the mechanics of simply putting a bin out'.77

The anti-litter campaign was delivered in three stages. The pre-Games element was entitled *Let's start training for the Games*; the Games time stage, *Clean up at the Games*; and the post-Games component, *Keep up the good work*.78 The programme's delivery involved a range of media, including television, radio, print and outdoor advertising, and sought to embrace all residents of Victorians, not simply those residing in the host city.79 Additionally, public relations activities were employed as an adjunct to media usage, which sometimes involved the use of high-profile athletes to launch clean-up events.80 Aiding the campaign's efforts were 16 regional waste management groups, Clean Up Australia, almost half (47%) of Victorian local councils and six environmental/community groups. These bodies performed various roles, including distributing/using materials prepared in support of the programme, conducting clean-up day events (81 were conducted involving 4300 volunteers) and/or assisting with media coverage of clean-up activities.81 Organisations involved in the campaign also had access to a detailed litter campaign resource kit.82

ECO-Buy Business. ECO-Buy Business commenced in July 2005 and was a partnership between OCGC, Sustainability Victoria and the Municipal Association of Victoria (the owners of ECO-Buy). It was also supported by the Australian Industry Group, the Victorian Employers' Chamber of Commerce and Industry, and the Industry Capability Network. The intent of this programme was to encourage, in particular, large businesses, to develop purchasing policies that would give preference to green products and services.83 In recognition of the importance of education in engaging businesses with the concept of 'green' purchasing, ECO-Buy Business acted to produce a guide on developing and implementing a green purchasing programme; conduct meetings designed to allow businesses to share their experiences regarding green purchasing; conduct exhibitions of green products and services; and deliver training sessions dealing with green purchasing.84

According to Adrienne Stephens, this programme was viewed as a legacy programme by the OCGC and was endorsed by major business associations.85 Efforts to stimulate membership in this initiative prior to and during the Games extended to feature articles in two magazines (*Waste Management* and *Environment and Contract Management in Practice*), industry presentations (14) and the conduct of a Green Purchasing Expo.86

Tree-planting programme. As part of an effort to make the Games a carbon-neutral event, the OCGC, in association with the Department of Sustainability and Environment, various catchment-management authorities, Greening Australia and community groups, aimed to plant up to 1 million trees.87 In order to achieve this goal, some 2000 volunteers were enlisted from organisations such as schools, scout groups and environmental organisations. According to Kendrea Pope, this programme was also viewed by the OCGC as a means of raising awareness about the issue of greenhouse gas abatement amongst those directly involved, as well as in the communities in the 15 regions in which tree planting took place.88 It should also be noted that a number of the tree-planting events were accompanied by presentations on indigenous species (flora and fauna), with one event involving a demonstration of direct seeding technology for local area landowners.89

GreenPower initiative. The GreenPower initiative was essentially a sponsorship arrangement between the OCGC and Origin Energy. This arrangement involved Origin Energy agreeing to sponsor the Games through a contra deal by which it supplied GreenPower (2.7 GWh) to all 16 Games venues located in Melbourne and in regional centres for the duration of the event. Additionally, Origin Energy provided 200 sqm of grid-connected solar photovoltaic panels for the new Olympic Stand entry at the Melbourne Cricket Ground.90 In return for its sponsorship of the event, Origin Energy was able to leverage its GreenPower actions associated with the MCG in a television campaign over the Games period.91 While the intent of this campaign was to increase the number of households signing up for this energy product, it can also be argued, from an ESD perspective, that Origin Energy's promotional actions served to further raise awareness throughout Victoria of 'green' energy in reducing greenhouse gas emissions. Indeed, the Victorian Minister for Environment, and the Minister for Energy Industries and Resources used the opportunity provided by Origin Energy's sponsorship of the Games to issue a press release highlighting this link.92

Grants programme. Commencing in mid-2005 and concluding at the end of March 2006, the OCGC operated a competitive grants programme called *Getting Involved*, which was open to all Victorian local councils. Eight of the projects awarded grants through this initiative were environment-based.93 Two of these programmes, the Yea Wetlands Boardwalk and the Bayside City Council's Indigenous Coastal Trail, involved a significant environmental interpretive component.94

In addition to competitive grants, OCGC also provided two one-off grants to MCC. The first of these was for $5 million to construct a 5-ha wetland area (Trin Warren Tam-boore) in the north-west section of Royal Park, Melbourne's largest park. This development was undertaken at the same time as the construction of the nearby MCG Athletes Village. The wetland was designed to treat and recycle storm water sourced from the Athletes Village, the surrounding area and Melbourne Park Zoo, using natural biological processes. In turn, this water was used to irrigate up to 20 ha of open space.95 From an ESD viewpoint, it is noteworthy that this artificial wetland was designed in such a way as to foster an appreciation of such environments through the inclusion of pathways, a bird hide and the onsite provision of printed information about the environmental attributes of the area.96 Insight Economics in its evaluation of the MCG was of the view that this initiative enhanced the awareness of wetland habitats and water recycling amongst Melbournians generally.97

The second grant received by MCC was for the development of a programme entitled *Savings in the City: Green Hotel Program*. This programme, which commenced in July 2005, sought to assist Melbourne hotels to reduce their use of energy and water and generation of solid waste.98 Significantly, from an ESD perspective, this initiative was underpinned by a number of publications and other resources developed to educate participating properties on how to engage with the programme. These publications were: an *EnergyWise Hotel Toolkit*, a *WasteWise Hotel Toolkit*; a benchmarking tool to gauge an individual hotel's efforts at minimising its energy, water usage and waste generation; and a *Green Hotels Water and WasteFact Sheet*.99Additionally, the City of Melbourne incorporated into its website a dedicated section from which these materials could be accessed or downloaded. By the end of the Games, this ongoing programme, according to its co-ordinator, Sheridan Blunt,100 had 15 participating properties.

3.4 Melbourne Commonwealth Games Corporation

M2006's role in the MCG's environmental programme was much more constrained than that of the OCGC. It was nonetheless classified as a focal organisation in this study due to its joint development of certain key aspects of this programme with the OCGC, most notably environmental procurement guidelines and waste-management processes. Additionally, M2006 was required to engage with the

MCG's environmental programme (as appropriate), which meant it acted to incorporate environmental procurement guidelines into tender documents; include environmental messages in its staff and volunteer training; and implement practices designed to minimise waste.101 M2006 also provided information about the environmental programme on its website, along with links to the event's tree-planting, education and 'Our River Our Games' programmes.102 Unlike the OCGC, M2006 did not seek to form collaborative stakeholder relationships for ESD purposes.

3.5 Department of Infrastructure

The need to transport large numbers of people to and from Games venues resulted in an existing programme, TravelSmart, being adapted, rebadged (TravelSmart Commonwealth Games) and implemented for this purpose by the Department of Infrastructure over the 2005–2006 period. This programme had two components: one focused upon communities immediately surrounding the three major Games venues (Melbourne Cricket Ground, Melbourne Sports and Aquatic Centre and Telstra Dome); and the other focused on large workplaces within the Melbourne CBD.103 In the case of the community programme, 11,000 households were targeted, while the participation goal for the workplace programme was 50 businesses, each employing 100 or more staff.104

The short-term goal of the TravelSmart Commonwealth Games programme was to convince people living or working in close proximity to Games venues to make use of alternative transport modes during the event and by so doing reduce greenhouse gas emissions and traffic congestion. Significantly from an ESD viewpoint, it was also intended that the programme would build a legacy of changed travel behaviour amongst its target groups.105

3.6 City of Melbourne

In addition to receiving grants from the OCGC to develop the Savings in the City: Green Hotel Program and to construct the Royal Park Wetlands, as previously noted, the City of Melbourne developed an independent relationship with Reiser Productions Pty Ltd. This relationship resulted in the co-production of a choir-based event entitled *Sing for Water*. This event, which was first produced for the 2002 Mayors Thames Festival in London, was incorporated into the Melbourne 2006 Commonwealth Games cultural programme and involved community choirs and schools performing songs about water.106The aim of this initiative was to further develop the Victorian community's awareness of the importance of clean water and sanitation in the context of developing countries, with a specific focus on this issue within Southeast Asia and the Pacific.107 A total of 27 community choirs took part in the main *Sing for Water* concert in Melbourne, while a separate event, *Children Sing for Water*, involved choirs from 12 primary schools.108 In total, some 5000 people attended these concerts, with profits donated to a clean water project in East Timor.109

3.7 Media

There were some 2713 accredited journalists at the MCG and its television viewing audience was claimed to be 1.5 billion.110The OCGC recognised the potential such coverage held to carry the event's environmental message and had developed a media and communication strategy specifically designed to create awareness of its environmental programme.111 A review of the media archives section of the M2006 and OCGC websites indicated that one means employed for this purpose was media releases. A total of 12 were identified, with the bulk of these (nine) emanating from the OCGC.112

While it is difficult to determine the extent of media coverage of the environmental programmes and initiatives associated with the Games, it is nonetheless possible to support the contention that coverage of this aspect of the event did indeed occur. In this regard, the media archives of the two largest Australian newspaper publishers, Fairfax Media Ltd and News Limited,113 were consulted commencing from the year the bid was won (1999) to the end of the year in which the event was delivered (2006).114 The total number of articles identified through this means that were linked to the MCG's environmental programme was 53 (or slightly less than 7 per year).

3.8 The Society for Sustainability and Environmental Engineering, the Environment Institute of Australia and the Waste Management Association of Australia

While some organisations within the business community acted to participate in, or support ESD initiatives (e.g. Green Trades Program and ECO-Buy Business), only a few were identified as active stakeholders in the ESD process. Specifically, these were the Society for Sustainability and Environmental Engineering, the Environment Institute of Australia and the Waste Management Association of Australia. These organisations, the first two being professional bodies and the last an industry association, jointly delivered a post-Games seminar entitled *Games Environment Report Card: Going for Green*. This seminar targeted professionals in the broad fields of environmental engineering, design and waste management, and sought to pass on some of what was learnt from the MCG to their respective constituencies.115

3.9 Factors Serving to Facilitate or Inhibit ESD in the Context of the MCG

3.9.1 Facilitating Factors

To assist it in developing its environment strategy, as noted earlier, the OCGC established an independent Advisory Committee to facilitate public input into the strategy-creation process –CGAC-E. This body acknowledged the opportunity provided by the Melbourne 2006 Commonwealth Games to demonstrate to the Victorian community how ecologically sustainable development could be applied in practice in areas such as waste, water and energy management. This acknowledgement, in turn, led to a number of recommendations (all but two of which were accepted by the Victorian Government) in its final report that are relevant to this study. Specifically, these recommendations relate to leveraging the opportunity presented by the event to raise awareness of environmental conservation and the environmental footprint associated with the lifestyles of individuals; generate long-term behavioural change in the areas of litter, public transport usage, energy and water usage; increase consumer awareness of environmentally friendly products and services; promote the purchase of sustainably produced goods; and enhance the understanding of practice in the environmental area by communicating the lessons and legacies of the Games' environmental programme. In addition to these more broadly based recommendations, the CGAE-E also saw the Games as an opportunity to educate private-sector organisations and their employees about the use of environmentally sustainable practices and technologies, believing this could be achieved by such means as showcasing, formal training and inclusion of environmental requirements in tender documents.116

It can also be reasonably argued that at the time of Victoria's bid for the CG in October 1999, the issue of sustainable development had been on the Australian public agenda for some time. For example, Australian governments (state and national) some seven years prior had collaborated to develop a *National Strategy for Ecologically Sustainable Development.*117In the context of Victoria alone, various developments can be identified in the period extending from the successful bid to the delivery of the Games that are also reflective of this engagement, including the conduct of a community summit in order to explore key themes in the states future development one of which was the environment118; the introduction of voluntary industry sustainability covenants in 2002119; introduction of a long-term water management strategy ('Our Water Our Future') in 2004120; and the release of the state's Greenhouse Strategy.121 It is noteworthy that these, and other such developments, were referred to by the CGAC-E in its report, arguably as a means of reminding government of its existing commitments in the area of sustainability in order to add weight to its recommendations.122

The previously cited actions of the Victorian Government reflect a desire on behalf of its community to engage with sustainable practices. Indeed, several interviewees were of the view that existing community expectations were such that there was no option but to engage with a sustainable development agenda through the MCG.123 Christine Forster summed up this situation, noting: 'I think the main factor that helped us along was that the overall general understanding in the community of sustainability is growing'.124

Arguably, the MCG was the first LSSE to take place in Australia that had access to a substantial stock of information concerning how such events can engage with an environmental agenda. In this regard, interviewees mentioned a number of past and upcoming events, specifically the 2006 Winter Olympic Games; 2006 FIFA World Cup; 2001 Goodwill Games; and the 2012 Olympic Games. In terms of the MCG's environmental programme, however, the SOG appears to have been the event that was drawn upon most for insights into this area. Batagol125 noted in this regard that while the amount of information available from other LSSEs as regards engaging with an environmental agenda increased as the MCG approached, the Sydney 2000 Olympic Games offered the most insights into this area up until the time of the event.

A number of interviewees acknowledged that the development of collaborative relationships by the OCGC with a variety of largely government and non-government organisations was crucial to the creation and delivery of MCG's environmental programme, and by inference, the ESD elements of it.126 A key aspect of their importance lies in the opportunity they provided to link with established initiatives (i.e. 'TravelSmart', anti-litter campaigns, ECO-Buy, Melbourne Water's community education programme, 'WasteWise'), which according to Batagol '... made it easier ... as ... things were already in train'.127 In some instances, additional funding provided by the OCGC to these established programmes allowed them to connect with larger or new audiences. Such was the case with TravelSmart, anti-litter efforts and ECO-Buy. By engaging organisations as stakeholders in its environmental programme, the OCGC was also able to access expertise in areas such as tree planting (Greening Australia); volunteer management (Conservation Volunteers Australia); litter reduction (VLAA); and education (Victorian Department of Education and Training). Grants and sponsorship, as noted earlier, were other means by which the OCGC acted to establish stakeholder relationships in its environmental programme.

3.10 Inhibiting Factors

An indication of the extent to which the environment was viewed as a central consideration in the planning and conduct of the MCG can be found in the *Commonwealth Games Arrangement Act 2001*, the main piece of Victorian Government legislation associated with the

MCG.128 The Act makes no mention of any environment-related deliverables or requirements associated with the Games. In acknowledgement of this omission, the Environment Liaison Office, a body representing peak environment groups in Victoria,129 wrote to the Victorian Government opposing the Act prior to its incorporation into state law. They argued, unsuccessfully, that the Act needed to be amended in order to incorporate ecologically sustainable development guidelines, and that an independent body should be established to perform a 'watchdog' role for the event's environmental dimension, as had been done in the context of the SOG – Green Games Watch 2000. Similarly, a rally outside the Victorian Parliament building by an alliance of environmental groups, which culminated in the presentation of a list of similar demands to the Minister for the Commonwealth Games, also failed to result in changes to the legislation.130

While absent from legislation, the Victorian Government (largely through the OCGC) did nonetheless engage with an environmental agenda through the MCG, with associated ESD elements, as has been discussed previously. While acknowledging this, Batagol131 argues that the environment was nevertheless an 'add-on' from the perspective of the OCGC rather than being integral to the event's management systems and associated decision-making processes. Marcus Godhino is supportive of this view, noting that from his interactions with Victorian Government public servants associated with the event, it was obvious that they were 'not trying to be Sydney ... and make as much of an effort as Sydney went to on the environment'.132 Cost may well have been a factor in this 'partial' adoption of a 'green' Games agenda. In this regard, Davidson133 notes that according to the successful construction tenderer (Village Park Consortium) for the Athletes Village, all bidders were asked midway through the tendering process to exclude environmental initiatives from their bids in order to reduce their quoted price. Subsequently, however, under pressure from the Melbourne and Moreland Councils (the local government areas in which the development was located), the Victorian Government agreed to fund the incorporation of environmental design elements into the Village.134 Issues associated with the State Government's concern over the cost of the Games are also evident in Christine Forster's communication that:

... a lot of sustainability initiatives that you might have liked to have done, like further schools programs which required much more upfront investment, just weren't possible in the context of the Games budget.135

Further evidence of the constrained emphasis on the 'green' elements of the MCG by the OCGC can be found in the level of media coverage the environmental dimension of the event received, even though a strategy had been developed for this purpose. In this regard, Batagol drew a parallel with the Sydney 2000 Olympic Games, noting in the context of the media that there was a 'real component of leveraging (the event) for sustainability purposes (by organisers)', something that she goes on to state was much less obvious in Melbourne.136 Kate McKeand was also of this view, observing that 'one of the things that we could have done more (of) is have more information in the media about the environment program'.137 The modest number of articles (as noted previously) published in Australia's two largest newspapers concerning the environmental aspects of the event adds further weight to this view.

The lack of media engagement around the environmental aspects of the Games resulted, according to Jenny Sams, in the broader community not making the connection between the environment and the MCG 18 months out from its commencement. It was this lack of profile of the environmental dimension of the MCG, she notes, that led to the decision to appoint an Environmental Ambassador whose brief was to address this issue.138 Still further evidence of this low-key approach to communicating the MCG's environmental dimensions can be found in the decision to omit any mention of this aspect of the event from the list of possible topics, of which there were 47, that could be covered in the event's observer programme.139 Finally, this limited effort to communicate the event's environmental component calls into question the degree to which the OCGC's media strategy linked to its environmental programmes and initiatives was either adequately resourced or implemented.

The limited time (three months) given to the CGEC-E to solicit community input into the environmental dimension of the MCG, and to undertake research before delivering its final report, can also be seen as symptomatic of the restricted emphasis the OCGC placed on the pursuit of an environmental agenda through the Games. Given this timeline, it is not surprising that a number of members of this committee believed that their ability to identify and suggest mechanisms for exploiting the full range of environmental opportunities, including those of an ESD nature, potentially associated with the event was severely constrained.140 The issue of lead times also affected other environmental elements of the event, with Rob Gell, the Environment Ambassador for the Melbourne 2006 Commonwealth Games, noting that by the time he was appointed (12 months out from the Games) it was largely too late to exploit the full potential of the event to carry an environmental message to the broader community.141

The bid criteria established by the CGF, to which the Australian Commonwealth Games Association (ACGA) and the Melbourne 2006 Commonwealth Games Bid Pty Ltd responded, necessitated only a basic acknowledgement of the environmental dimension of the MCG. This is reflected in the bid document for the event in which the component dealing with the environment was included in the section headed *Additional Information*. Information included here in relation to the environment was limited to some 300 words which comprised general statements to the effect that the event would employ world's best practice in environmental management; utilise environmentally sensitive approaches in the areas of transport, energy and waste management; engage in recycling; and construct an energy-efficient Athletes Village.142 The brevity of this section raises the issue of the degree to which the CGF (at the time of the MCG) sought to embrace environmental matters in selecting CG host cities. Additionally, given, according to Batagol, that the CGAC-E was unaware of the contents of this section of the bid document, its influence on subsequent actions taken by organisers in the environmental area would seem to be limited.143 Further, it is worthy of note that the ACGA website's listing of sections contained in the MCG bid document omitted any mention of the environment. Given this apparent restricted focus on the environment by CGF and its ACGA affiliate, it is perhaps not surprising that the environmental programme for the MCG, as Pope notes, was 'very much a government-driven agenda as opposed to a CGF-driven requirement'.144

The extent to which a more engaged CGF and/or ACGA in the context of the MCG's environmental programme would have influenced ESD outcomes cannot be known. Nonetheless, the question of the effect of a more involved event 'owner' and local affiliate must be asked.

It would also appear that M2006 had only a limited interest in the environmental agenda established for the MCG. In this regard, Batagol notes that 'we worked hard at making sure that M2006 built into their planning for the Games the environment features that we wanted and needed' and asserts that the management of M2006 'simply did not see sustainability matters as being connected to their task of delivering the Games'.145 Given this lack of interest in environmental considerations as the event approached, the OCGC needed, according to Callaghan, to be increasingly engaged with ensuring M2006 was involved with the MCG's environmental agenda.146

Evaluation of the MCG's environment programme took the form of a triple bottom line post-Games analysis which embraced the social, economic and environmental aspects of the event (including its ESD components) conducted by Insight Economics; a post-Games audit of the event's final environment report card by the Victorian Commissioner for Environmental Sustainability; and contracted consultant reviews of selected environmental programmes, specifically the Games' litter campaign, tree-planting programme and TravelSmart initiative.147 Missing from this 'mix' of reviewers is the 'arms length' body suggested, as noted earlier, by the Environment Liaison Office and the CGAC-E. Such a body, according to the CGAC-E, would have taken the form of an independently chaired Environment Stakeholder Reference Group comprising representatives from peak environmental organisations, industry associations, local government, public transport user groups, resident groups and unions. This group, they envisaged, would have acted to produce quarterly publicly available reports on the environmental aspects of the Games and provide ongoing advice to the Minister for Commonwealth Games on the implementation of the Government's response to the CGAC-E's report.148

While the OCGC did nonetheless establish an ESF, its form differed from that suggested by the CGAC-E, with its members being drawn largely from government entities. One possible interpretation of constraining membership of this body is that the Victorian Government wished to reduce broader community engagement in the MCG's environmental programme, as well as any potential controversies and resulting media coverage. From an ESD perspective, this lack of engagement arguably inhibited community discussion and debate around environmental matters linked to the Games.

While the OCGC may have sought to 'control' the review process for the MCG's environmental programme, it should nonetheless be noted that environmental advocacy groups such as Greenpeace chose not to become involved with the event. In this respect, the MCG differed significantly from the SOG. Batagol notes that '... you actually had green groups driving agendas very, very strongly (in Sydney). In Melbourne they just weren't there'.149 There was, as stated earlier, some measure of interest from these groups initially, as they lobbied to alter legislation (*Commonwealth Games Arrangement Act*) associated with the event to encompass ecologically sustainable development guidelines; however, once this was rejected by the State Government there appears to have been little substantial follow-up.

4. DISCUSSION

The MCG case used in this exploratory study has identified a number of practices with the potential to generate ESD outcomes for an LSSE host community. Some of these programmes/activities were found to be limited to specific groups, while others had the capacity to impact the community more broadly. Of particular note were use of environment-linked procurement guidelines; incorporation of environmental components into staff, contractor and volunteer training; production/commissioning of publicly available environment-related reports and other publications; media engagement with the event's environmental programme; provision of financial support for environment-linked projects with educational elements; inclusion of environmental material within school curricula; event-linked environmental campaigns, awareness raising and behaviour change programmes; the showcasing of environmentally friendly construction and design; and the transfer of environment-related information by such means as seminars/workshops.

Evidence as to the effect of these programmes/activities on host community ESD is limited; however, it is noteworthy that a post-

Games attitudinal study found that 60% of Victorians were more willing to act in ways that were of benefit to the environment as a result of the event. Additionally, 47% of respondents to this survey stated they were more likely to use public transport as a direct result of focus placed upon it during the event.150 These findings were attributed to the inclusion of environmental elements in curriculum material prepared for the event's schools programme; the participation of community volunteers in a range of environmental initiatives (e.g. tree-planting days); public communications programmes in the areas of public transport use and litter; and public exposure to environmental messages linked to the Games. Arguably, the blending back into the community of some 15,000 volunteers, along with OCGC and M2006 staff, whose training had incorporated environmental elements, may have also been a further contributing factor here.

This enquiry has also highlighted the significance of government, and of government commitment, in the process of leveraging LSSEs for ESD purposes. Additionally, it has identified other factors that potentially contribute to the use of these types of events in order to generate ESD outcomes, specifically a climate of expectation within the host community; access to information as regards how to engage with a sustainability agenda (inclusive of ESD) through an LSSE; and the value of focal organisations collaborating with other organisations and groups in order to expand the range and impact of ESD-linked initiatives.

A range of factors were also found to be in evidence that impacted the degree to which ESD outcomes emerged from the event. Of particular note in this regard were the 'second-tier' status given by the Victorian State Government and M2006 to the event's environmental programme. This was evident from the short period of time given to the CGAC-E to engage the host community regarding the leveraging of the MCG for sustainable development purposes; the decision not to create an 'arms length' body to critique and provide ongoing input into the event's environmental programme; and the lack of any mention of the environment in the MCG's enabling legislation. Other potentially limiting factors were found to be the absence of any environmental focus by the event's 'owner' (the CGF) and local affiliate (the ACGA); and the failure by the MCG's focal organisations to actively engage the media around the event's environmental programme. The decision by independent environmental groups, such as Greenpeace and WWF, not to seek involvement in the event compounded this lack of media focus.

When viewed from the perspective of the definition of ESD stated at the outset of this paper, this study would seem to suggest that the impact of LSSEs in host communities may be constrained to the more foundational elements of ESD – awareness raising, attitudinal and behaviour change, and the development of an enhanced capacity to engage in sustainable practices. While such outcomes are important from an ESD perspective, by themselves they do little to develop the skills and knowledge conducive to the broader social, political and economic changes that are key to sustainable development. This being the case, host communities seeking to leverage LSSEs for ESD purposes are perhaps advised to view them more as 'one-off' opportunities to progress aspects of their broader efforts at ESD rather than as drivers of fundamental change in dealing with the challenge posed by sustainable development.

5. FUTURE RESEARCH

This study, while providing meaningful insights into the leveraging of LSSEs for ESD purposes, is limited by its focus on a single event. Deeper insights into this issue would result from examining other LSSEs that have purposely sought to embrace a sustainability agenda inclusive of ESD. Additionally, emerging from this enquiry are specific challenges faced by host communities seeking to use such events as vehicles through which to progress their ESD efforts that would benefit from further study. These include how to maximise the potential opportunities LSSEs present for ESD, particularly given their limited life, and related to this, how best to integrate ESD considerations within the broader event legacy planning process.

NOTES

1. UNESCO, "Education for Sustainable Development," 2.

2. See, for example, DEHA, *Living Sustainably*.

3. See, for example, NSWCEE, *Learning for Sustainability*; DSE, *Learning to Live Sustainably*.

4. Harris, "Educating Communities for a Sustainable Future."

5. UN, *United Nations Conference*, Chapter 36.1.

6. See, for example, SDEP, *Learning to Last*; Heck, "The State of Environmental Education"; NZPCE, *Sea Change*; DEHA, *Living Sustainably*; ARIES, *Education for Sustainability*; CMEC, *Education for Sustainable Development*; and UNESCO, "UN Decade of Education."

7. Fien, *Education for Sustainability*, 6.

8. Lenskyj, "Sport and Corporate Environmentalism."

9. Mallens, Stevens and Adams, "A Content Analysis."

10. UNEP, *Long-Term Strategy*.

11. Ibid., 1.

12. CECM, *Recommendation 17 of the Committee*.

13. Green Cross Spain, *Green Charter*.

14. Global Sports Alliance, "Home Page."

15. WWF, *Environmental Assessment*; WWF, *WWF Works with Olympic Organizers*; BioRegional and WWF, "Towards a One Planet"; Greenpeace, *Greenpeace Olympic Environmental Guidelines*; and Greenpeace, *The Beijing 2008 Games.*

16. Ottesen, "The Olympic Vision."

17. IOC, "Sport and Environment Commission."

18. UNEP, "London 2012 Will Leave a Lasting Legacy."

19. FIFA, *Green Goal Programme.*

20. Global Sports Alliance, "Global Forum for Sports."

21. See note 4 above.

22. The Commonwealth is a voluntary association of 71 countries with a collective population of over 2 billion people (seeCGF, *Constitution, Games Management Protocols*; Insight Economics, "Triple Bottom Line Assessment").

23. CGF is an arm of the Commonwealth. The CGF's role is to direct the development of the Commonwealth Games, which is conducted on a four-yearly cycle. Additionally, it has the broader role of assisting with the development of sport throughout the Commonwealth (see CGF, *Constitution, Games Management Protocols*).

24. ACGA and Melbourne 2006 Commonwealth Games Bid Pty Ltd, "Melbourne's Bid."

25. CGF, *2018 Commonwealth Games Candidate.*

26. MCGOC, *Post Games Final Report.*

27. UNEP, "The 2010 Commonwealth Games."

28. Glasgow City Council, *Glasgow 2014 Commonwealth Games.*

29. Stake, "Case Studies"; Stenhouse, "Case Study Methods."

30. OECD, *Education and Learning.*

31. Gillham, *Case Study Research Methods.*

32. Neuman, *Basics of Social Research.*

33. Patton, *Qualitative Research and Evaluation.*

34. Stoecker, cited in Neuman, *Basics of Social Research.*

35. For example, Platt, "What Can Case Studies Do?"; Stake, "Qualitative Case Studies"; and Patton, *Qualitative Research and Evaluation.*

36. Established in February 2002, the OCGC was located within the Victorian Government's Department for Victorian Communities. The OCGC performed a broad strategic planning and management role, and was responsible for the formulation of government policy in support of the delivery of the MCG; coordination of all bodies associated with the event; infrastructure development; coordinating public domain events; traffic and transport arrangements; and contract security (see TTA and ARTV, "Scoping Report"; Australian Government, "Victorian Government").

37. OCGC, *2006 Melbourne Commonwealth Games: The Environmental,* 7.

38. OCGC, "Environment Program."

39. OCGC, *Melbourne 2006 Commonwealth Games.*

40. OCGC, "About Us."

41. Victorian Parliament, *Commonwealth Games Arrangements Act,* s.1A, ss. 4B.

42. OCGC, *Commonwealth Games Environment Procurement.*

43. Ibid.

44. Insight Economics, "Triple Bottom Line Assessment."

45. OCGC, *Melbourne 2006 XVIII Commonwealth Games.*

46. Buxton, personal communication, November 6, 2006.

47. The OCGC established independent Advisory Committee whose role was to facilitate public input into the MCG's environmental strategy. This body had five members drawn from government bodies associated with natural resource management (three), non-government environment groups (one) and universities (one); OCGC, *Melbourne 2006 XVIII Commonwealth Games.*

48. OCGC, *2006 Melbourne Commonwealth Games: Government Response.*

49. Gell, personal communication, April 20, 2007.

50. VDET, *Commonwealth Games Education Project.*

51. OCGC, *Melbourne Commonwealth Games Education Program.*

52. OCGC, *CG Reps Gamezine.*

53. See note 49 above.

54. VAEE, *Newsletter of Victorian Association.*

55. See note 38 above.

56. McKeand, personal communication, April 19, 2007.

57. Ecocycle Victoria, *Waste Water and Sustainable Schools.*

58. M2006, "Getting Involved."

59. Sustainability Victoria, "Sustainability: Green Plumbers."

60. See note 38 above.

61. Sustainability Victoria, "Sustainability: Waste Wise Catering."

62. See note 42 above.

63. Major Projects Victoria, "Parkville Gardens."

64. DVC, *Annual Report 2003–04*.

65. See note 42 above.

66. Ibid.

67. Insight Economics, "Triple Bottom Line Assessment."

68. Ibid.; see also note 42 above.

69. OCGC, *Our River, Our Games*.

70. See note 42 above.

71. See note 52 above.

72. See note 42 above.

73. McDonald, personal communication, April 26, 2007.

74. See note 42 above.

75. Cosson, personal communication, May 5, 2007.

76. VLAA, "A Litter-Free Commonwealth Games."

77. Batagol, personal communication, March 16, 2007.

78. See note 42 above.

79. Cossen, personal communication, May 5, 2007.

80. Sustainability Victoria, *Sustainability Victoria Litter Prevention*.

81. See note 42 above.

82. Community Change, *Evaluation of the Commonwealth Games*.

83. See note 42 above.

84. ECO-Buy, "About ECO-Buy."

85. Stephens, personal communication, May 29, 2007.

86. See note 42 above.

87. See note 67 above.

88. Pope, personal communication, May 4, 2007.

89. See note 54 above.

90. OCGC, "Green Venues."

91. Origin Energy, "Origin Energy Promotes a Greener Games."

92. Minister for Environment and Minister for Energy Industries and Resources, "Origin Energy Provides Green Power."

93. See note 42 above.

94. Yea Wetlands Committee of Management, *Yea Wetlands Walk*; Bayside City Council, "Indigenous Coastal Trail."

95. See note 42 above.

96. City of Melbourne, "A Self-Guided Walk."

97. See note 42 above.

98. Ibid.

99. City of Melbourne, "Savings in the City."

100. Blunt, personal communication, April 18, 2007.

101. See note 42 above.

102. M2006, "Getting Involved."

103. TravelSmart, "TravelSmart Commonwealth Games."

104. TravelSmart, *TravelSmart Communities*; D. Meiklejohn, personal communication, March 6, 2007.

105. See note 42 above.

106. Sing for Water Australia, "Home."

107. See note 42 above.

108. Ibid.

109. See note 104 above.

110. See notes 42 above.

111. See note 42 above.

112. OCGC, "News and Publications"; M2006, "Media Release Archives."

113. Fairfax Digital, *News Store*; News Limited, "Newstext."

114. The review process involved a search of the online newspaper databases of the News Limited and Fairfax corporations using the broad headings of Melbourne Commonwealth Games, Melbourne 2006 Games and Commonwealth Games. A further screening process then took place to identify those that dealt with one or more aspects of the environmental dimensions of the event.

115. Engineers Australia, *Games Environmental Report Card.*

116. CGAC-E, *Report to the Minister.*

117. ESD Steering Committee, *National Strategy.*

118. DPC, *Growing Victoria Together.*

119. See note 42 above.

120. Victorian Government, *Our Water Our Future.*

121. DSE, *Victorian Government Greenhouse Strategy.*

122. See note 114 above.

123. Pope, personal communication, May 4, 2007; McKeand, personal communication, April 19, 2007.

124. Forster, personal communication, March 19, 2007.

125. Batagol, personal communication, April 16, 2007.

126. McKeand, personal communication, April 19, 2007; Gonzalvez, personal communication, June 6, 2007; Callaghan, personal communication, June 6, 2007; Blunt, personal communication, April 18, 2007; and Cosson, personal communication, May 5, 2007.

127. See note 125 above.

128. See note 39 above.

129. At the time of the MCG, these bodies were the Australian Conservation Foundation, Friends of the Earth, Environment Victoria, Victorian National Parks Association and the Wilderness Society Victoria; see Royal Park Protection Group, *Annual Report 2001.*

130. Royal Park Protection Group, *Annual Report 2001.*

131. See note 77 above.

132. Godhino, personal communication, November 7, 2006.

133. Davidson, "Can We Afford the Parkville?"

134. Millar, "Plans for 'Greener' Games."

135. See note 124 above.

136. Batagol, personal communication, March 3, 2007.

137. See note 54 above.

138. Sams, personal communication, November 11, 2007.

139. DVC, *Melbourne 2006 Commonwealth Games Observers Program.*

140. Buxton, personal communication, November 6, 2006; Read, personal communication, April 20, 2007; Godhino, personal communication, November 7, 2007; and Foster, personal communication, March 19, 2007.

141. See note 49 above.

142. See note 24 above.

143. See note 77 above.

144. Pope, personal communication, May 5, 2007.

145. See note 77 above.

146. Callaghan, personal communication, June 6, 2007.

147. See note 42; CES, *Independent Assessment.*

148. See note 114 above.

149. See note 77 above.

150. See note 42 above.

REFERENCES

1 ACGA, and Melbourne 2006 Commonwealth Games Bid Pty Ltd.Melbourne's Bid for the XVIII Commonwealth Games. Melbourne: ACGA & Melbourne 2006 Commonwealth Games Bid Pty Ltd, 1999.

2 Australian Government. "Victorian Government: Office of Commonwealth Games Co-Ordination." Australian Government, n.d.

3 Australian Research Institute in Education for Sustainability (ARIES). Education for Sustainability: The Role of Education in Engaging and Equipping People for Change. Sydney: Macquarie University, 2009.

4 Bayside City Council. "Indigenous Coastal Trail: Council News Column." Indigenous Coastal Trail: Council News Column." Bayside City Council, 2005.

5 BioRegional, and WWF. "Towards a One Planet Olympics Revisited.". http://tinyurl.com/ce7nnrn2012.

6 CGAC. Report to the Minister for Commonwealth Games on the Melbourne 2006 Commonwealth Games. Melbourne: CGAC-E, 2002. http://www.ausport.gov.au/fulltext/ 2002/vic/CG_advisory_committeereport.asp.

7 CGF. 2018 Commonwealth Games Candidate City Manual. London: CGF, May 2011.http://www.thecgf.com/ games/2018/2018-Candidate-City-Manual.pdf.

8 CGF. "Constitution, Games Management Protocols, Regulations and Games Conduct." London: CGF, 2005.http:// www.thecgf.com/ about/constitution.pdf.

9 City of Melbourne. "Savings in the City: Green Hotels Program." City of Melbourne, Melbourne, n.d. Accessed April 1, 2008.http://www.melbourne.vic.gov.au/info.cfm?top = 218&pg = 2546.

10 City of Melbourne. "A Self-Guided Walk in a City Wetland." City of Melbourne, Melbourne, n.d. Accessed May 9, 2007.http://www.thatsmelbourne.com.au/tmcarsrc/wetlands/tour.pdf.

11 City of Melbourne Council (CMC). City of Melbourne Annual Report 2006–07. Melbourne: CMC, 2006.http://www.melbourne.vic.gov.au/About Council/Plansand Publications/AnnualReport/Documents/Annual_Report_2006_07.doc#_Toc178585657.

12 Clean Up the World. "The Clean Up Story." Clean Up the World, Sydney, n.d. Accessed April 1, 2008.http://www. cleanup.org.au/en/About/the-clean-up-story.html.

13 Commissioner of Environmental Sustainability (CES). Independent Assessment of the Green Games. Melbourne: CES, n.d. Accessed April 10, 2008.http://www. ces.vic.gov. au/CES/wcmn301.nsf/obj/publications/$file/Commonwealth+Games+Report+Card.pdf.

14 Community Change. Evaluation of the Commonwealth Games Litter Campaign. Melbourne: Community Change, 2006.

15 Conservation Volunteers. "About Us." Conservation Volunteers, Ballarat, n.d. Accessed August 13, 2008.http://www.conservationvolunteers.com.au/about-us.

16 Council of Europe: Committee of Ministers (CECM). Recommendation 17 of the Committee of Ministers to Member States on the Code of Sustainability in Sport: A Partnership Between Sport and the Environment. Strasbourg: CECM, 2000.https://wcd.coe.int/ViewDoc-jsp?id =

371057&BackColorInternet = 9999CC&BackColorIntranet = FFBB55&BackColorLogged = FFAC75.

17 Council of Ministers of Education, Canada (CMEC). Education for Sustainable Development in Canadian Faculties of Education. Toronto: CMEC, 2012.http://www.cmec.ca/ Publications/ Lists/ Publications/Attachments/279/ESD_Dean_reportEN.pdf.

18 Davidson, Kenneth. "Can We Afford the Parkville Games Village?" The Age, June 16, 2003.http://www.theage. com.au/articles/ 2003/06/15/1055615673360.html.

19 Department of Education and Early Childhood Development (DEECD). "About Us." DEECD, Melbourne, n.d. Accessed April 1, 2008. http://www.education.vic.gov.au/ about/department/ Pages/default.aspx.

20 Department of Premier and Cabinet (DPC). Growing Victoria Together. Melbourne: DPC, 2001.http://www.dpc.vic. gov.au/ CA256D800027B102/Lookup/GVTIIBooklet/$file/growing_vic_together%20final%20report.pdf.

21 Department of Sustainability and Environment (DSE). Victorian Government Greenhouse Strategy Progress Report 2003. Melbourne: DSE, 2003.http://www. climatechange.vic. gov.au/ CA256F310024B628/0/4C85C4FD7AFC8290CA2573830081E6C1/$File/VGS+Progress+Report+2003.pdf.

22 Department of the Environment, Heritage and the Arts (DEHA). Living Sustainably: The Australian Government's National Action Plan for Education for Sustainability. Canberra: Commonwealth of Australia, 2009.

23 Department of Victorian Communities (DVC). Annual Report 2003–04. Melbourne: DVC, 2004.http://www.dpcd.vic. gov.au/ Web14/dvc/rwpgslib.nsf/GraphicFiles/Corporate+2/$file/Corporate2.pdf.

24 DSE. Learning to Live Sustainably: Victoria's Approach to Learning-Based Change for Environmental Sustainability. Melbourne: DSE, 2005. http://www.dpi.vic.gov. au/ CA256F310024-B628/0/2BDA5FDE21AB1A19CA2570910011 9B24/ $File/content-v-6.pdf.

25 DSE. "Sustainable Household Challenge." DSE, Melbourne, 2007. http://www.dse.vic.gov.au/thesustainables/teachers.htm.

26 DVC. Melbourne 2006 Commonwealth Games Observers Program. Melbourne: DVC, n.d.

27 ECO-Buy. "About ECO-Buy.". ECO-Buy, Melbourne, 2006. http://www.ecobuy.org.au/director/about.cfm.

28 Ecocycle Victoria. Waste Wise and Sustainable Schools Awards Supported by Sustainability Victoria. Melbourne: Ecocycle Victoria, 2005.

29 Engineers Australia. Games Environment Report Card: Going for Green. Melbourne: Engineers Australia, 2006.

30 Environmental Protection Authority (EPA). "About EPA Victoria." EPA, Melbourne, 2007.http://www.epa.vic. gov.au/about_us/ default.asp.

31 ESD Steering Committee. National Strategy for Ecologically Sustainable Development. Canberra: AGPS, 1992.

32 Fairfax Digital. News Store. Sydney: Fairfax Digital, 2007. https://www.google.com.au/#q = fairfax+news+store.

33 Fien, John. Education for Sustainability: Reorientating Australian Schools for a Sustainable Future. Melbourne: The Australian Conservation Foundation, 2001.

34 FIFA. Green Goal Programme. Zurich: FIFA, n.d. Accessed September 7,

2007.http://www.fifa.com/worldcup/archive/southafrica2010/organisation/greengoal/programme.html.

35 Gillham, Bill. Case Study Research Methods. London: Continuum, 2000.

36 Glasgow City Council. Glasgow 2014 Commonwealth Games Sustainability Plan. Glasgow: Glasgow City Council, 2010. https://www.glasgow.gov.uk/CHttpHandler.ashx?id = 8465&p = 0.

37 Global Sports Alliance. "Global Forum for Sports and the Environment. 4th East Asian Games." GSA, Tokyo, n.d.

38 Global Sports Alliance. "Home Page." GSA, Tokyo, n.d. Accessed October 5, 2007. http://www.gsa-world.org/english/gsa/.

39 Green Cross Spain. Green Charter of Spanish Sport. Madrid: GCS, 2007. Accessed April 3, 2009.http://www.deportesostenible.es/doc/Internacionales/Green_Charter.pdf.

40 Greening Australia, and Green Corps. Green Corps and the Environment. Brisbane: Greening Australia, n.d. Accessed February 8, 2010. http://greencorps.greeningaustralia. org.au/category.php?id = 68.

41 Greenpeace. The Beijing 2008 Games Environmental Performance Evaluation. Sydney: Greenpeace, 2008.http://www.greenpeace.org/raw/content/china/en/press/-reports/green.pdf.

42 Greenpeace. Greenpeace Olympic Environmental Guidelines: A Guide to Sustainable Events. Sydney: Greenpeace, 2000. http://www.greenpeace.org.au/archives/olympics/reports/new guidelines.pdf.

43 Harris, Robert James. "Educating Communities for a Sustainable Future: Do Large Scale Sporting Events Have a Role?." Doctorate

of Education dissertation, University of Technology, Sydney, 2010.http://www.business.uts.edu.au/acem/pdfs/LargeScaleSportingEventsDevelopment.pdf.

44 Heck, Debbie. "The State of Environmental Education in the Australian School Curriculum." Australian Journal of Environmental Education19 (2003): 115–124.

45 Insight Economics. "Triple Bottom Line Assessment of the XVIII Commonwealth Games." Insight Economics, Melbourne, 2006.

46 IOC. "Sport and Environment Commission." IOC, Lausanne, n.d. http://www.olympic.org/sport-environment-commission.

47 Lenskyj, Helen Jefferson. "Sport and Corporate Environmentalism: The Case of the 2000 Olympics." International Review for the Sociology of Sport33, no. 4 (1998): 341–354.

48 M2006. "Getting Involved: Environment Program." M2006, Melbourne, 2006.http://www.melbourne2006. com.au/Getting+Involved/Environmental+Program/.

49 M2006. "Media Release Archives." M2006, Melbourne, 2006.http://www.melbourne2006.com.au/Media+Centre/Media+Release+Archive/.

50 Major Projects Victoria. "About Major Projects Victoria." MPV, Melbourne, 2008. Accessed April 6, 2008.http:// www.majorprojects.vic.gov.au/web3/majorproj.nsf/headingpagesdisplay/About+Major+Projects+Victoria?OpenDocument&Expand = 2&.

51 Major Projects Victoria. "Parkville Gardens." MPV, Melbourne, 2008. Accessed April 6, 2008.http://www.majorprojects.vic.gov.au/Web3/majorproj.nsf/AllDocs/CFAA7E0ACB7957D2CA256C24001A6450?OpenDocument&Expand = 3.15&.

52 Mallens, Cheryl, JulieStevens, and Lorne J.Adams. "A Content Analysis of Environmental Sustainability Research in Sport Management Literature." Journal of Sport Management25, no. 3 (2011): 240–256.

53 Manchester Commonwealth Games Organising Committee (MCGOC). Post Games Final Report. Manchester: MCGOC, 2002.

54 Melbourne Water. Melbourne Water: Sustainable Water a Strategic Framework. Melbourne: Melbourne Water, 2007.

55 Millar, Royce. "Plans for 'Greener' Games Village." The Age, August 6, 2003.www.theage.com.au/articles/ 2003/08/05/ 1060064182029.html.

56 Minister for Environment, and the Minister for Energy Industries and Resources. "Origin Energy Provides Green Power for the Games." Media release, February 24, 2006.http:// www.sustainability.vic.gov.au/resources/documents/060225_Greenpower_Games.pdf.

57 Neuman, Laurence W.Basics of Social Research: Quantitative and Qualitative Approaches. Needham Heights, MA: Allyn and Bacon, 2003.

58 New South Wales Council on Environmental Education (NSWCEE). Learning for Sustainability: NSW Environmental Education Plan 2007–10. Sydney: NSWCEE, 2006.

59 New Zealand Parliamentary Commissioner for the Environment (NZPCE). Sea Change: Learning and Education for Sustainability. Wellington: NZPCE, 2004. http://www.aries.mq.edu.au/ portal/ pdfs/UnderstandingEfS/SeeChange.pdf.

60 News Limited. "Newstext." News Limited, 2007. http:// www.newstext.com.au.

61 OCGC. 2006 Melbourne Commonwealth Games: The Environmental Framework. Melbourne: OCGC, 2002.http://www.ausport.gov.au/fulltext/2002/vic/CG_environment_framework.pdf.

62 OCGC. "2006 Melbourne Commonwealth Games: Government Response to Commonwealth Games Advisory Committee (Environment) Recommendations." Melbourne: OCGC, 2003. http://catalogue.ausport.gov.au/fulltext/ 2003/vic/CG-_govt_response.pdf.

63 OCGC. "About Us." OCGC, Melbourne, 2006.

64 OCGC. CG Reps Gamezine. Melbourne: OCGC, 2005. http://education.melbourne2006.com.au/?s = cgreps.

65 OCGC. Commonwealth Games Environment Procurement Policy. Melbourne: OCGC, n.d.

66 OCGC. "Environment Program." OCGC, Melbourne, n.d. Accessed March 7, 2007.http://www.melbourne2006. com.au/Getting+Involved/Environmental+Program/.

67 OCGC. "Green Venues." OCGC, Melbourne, n.d.

68 OCGC. Melbourne 2006 XVIII Commonwealth Games Environment Report Card. Melbourne: OCGC, 2006.

69 OCGC. Melbourne 2006 Commonwealth Games: The Environment Strategy. Melbourne: OCGC, n.d.

70 OCGC. Melbourne Commonwealth Games Education Program: Let the Games Begin – Student Resource: Environmental Strategy. Melbourne: OCGC, 2005.

71 OCGC. "News and Publications." OCGC, Melbourne, 2006.

72 OCGC. Our River, Our Games. Melbourne: OCGC, 2006.

73 Organisation for Economic Cooperation and Development (OECD). Education and Learning for Sustainable Consumption. Paris: Environment Directorate, Centre for Educational Research and Innovation, OECD, 1999.

74 Origin Energy. "Origin Energy Promotes a Greener Games." Origin Energy, Sydney, 2006.http://www. originenergy. com.au/ news/news_detail.php?pageid = 82&newsid = 654.

75 Ottesen, Peter. "The Olympic Vision." In The Green Games: A Golden Opportunity: Proceedings of a Conference Organised by the Centre for Olympic Studies, edited by RichardCashman, and AnthonyHughes, 32–39. Sydney: Centre for Olympic Studies, University of New South Wales, 1998.

76 Parks Victoria. Annual Report 2006–07. Melbourne: Parks Victoria, 2007.http://www.parkweb.vic. gov.au/ resources/ 03_ 2028.pdf.

77 Patton, Michael. Qualitative Research and Evaluation Methods., 3rd ed.Thousand Oaks, CA: Sage, 2002.

78 Platt, Jennifer. "What Can Case Studies Do?" In Qualitative Research, Vol. 1, edited by AlanBryman, and Robert G.Burgess, 160–179. London: Sage, 1999.

79 Royal Park Protection Group. Annual Report 2001. Melbourne: RPPG, 2001.http://www.royalparkprotect. org.au/Annual% 20Reports/AR2001.pdf.

80 Sing for Water Australia. "Home." Sing for Water Australia, n.d. Accessed June 9, 2007.http://www.singforwateraustralia.com/.

81 Stake, Robert E. "Case Studies." In Handbook of Qualitative Research, edited by Norman K.Denzin, and Yvonna S.Lincoln, 236–247. Thousand Oaks, CA: Sage, 1994.

82 Stake, Robert E. "Qualitative Case Studies." In The Sage Handbook of Qualitative Research, edited by Norman K.Denzin, and Yvonna S.Lincoln, 443–466. 3rd ed.Thousand Oaks, CA: Sage, 2005.

83 Stenhouse, Lawrence. "Case Study Methods." In International Encyclopaedia of Education, edited by NevilePostlethwaite, and TorsunHusan, 645–650. Oxford: Pergamon, 1985.

84 Sustainability Victoria. Our Business: 2007–2010. Melbourne: Sustainability Victoria, n.d. Accessed February 21, 2008. http://www.sustainability.vic.gov.au/resources/documents/SV4949_bus_plan_FA.pdf.

85 Sustainability Victoria. "Sustainability: Green Plumbers Training Sessions." Sustainability Victoria, Melbourne, 2006.www.sustainability.vic.gov.au/www/html/2023-greenplumbers-training-session.

86 Sustainability Victoria. "Sustainability: Waste Wise Catering Train-the-Trainer Resource." Sustainability Victoria, Melbourne, 2006. www.sustainability.vic.gov.au/www/html/1249-waste-wise-catering-train-the-tr.

87 Sustainability Victoria. "Sustainability Victoria Litter Prevention Grants 2005–06 Case Study: "Lets Start Training for the Games" Regional Litter Campaign." Melbourne: Sustainability Victoria, 2006.http://www.sustainability.vic.gov.au/resources/documents/Litter_Grant_Case_Study_Barwon_-_for_web.pdf.

88 Sustainable Development Education Panel (SDEP). Learning to Last: The Government's Sustainable Development Education Strategy for England. London: SDEP, 2003. http://www.defra.gov.uk/environment/sustainable/educPanel/ pdf/ sdeduc_-draftstrat.pdf.

89 Tourism Training Australia, and Arts and Recreation Training Victoria (TTA and ARTV). "Scoping Report: Strategic Training Issues and Recommendations for the 2006 Commonwealth Games." TTA and ARTV, 2002.

90 TravelSmart. "TravelSmart Commonwealth Games." TravelSmart, Melbourne, 2007.

91 TravelSmart. TravelSmart Communities: Commonwealth Games. Melbourne: TravelSmart, 2006.

92 UN. United Nations Conference on Environment and Development: Agenda. New York: UN, 1992.http:// www.un.org/esa/sustdev/documents/agenda21-/english/agenda21toc.htm.

93 UNEP. "The 2010 Commonwealth Games Go Green." Media release, February 17, 2010. UNEP, Nairobi. www. unep.org/Documents.Multilingual/Default.asp?DocumentID = 612&ArticleID = 6463&l = en.

94 UNEP. "London 2012 Will Leave a Lasting Legacy for the UK and the Olympic Movement: UNEP Executive Director." Media release. UNEP, Nairobi, 2012. http://www.unep.org/ Documents.Multilingual/Default.asp?-DocumentID = 2691&ArticleID = 9239&l = en.

95 UNEP. "Long-Term Strategy for Sport and the Environment of the United Nations Environment Program.". Nairobi: UNEP, 2003. www.unep.org/sport_env/Documents/Michezo_strategy.doc.

96 UNESCO. "Education for Sustainable Development United Nations Decade (2005–2014): Vision & Definition of ESD." UNESCO, Paris, n.d. Accessed August 10, 2007. http://portal.unesco.org/-education/-en/ev.php-URL_ID = 27279&URL_-DO = DO_TOPIC&URL_-SECTION = 201.html.

97 UNESCO. "UN Decade of Education for Sustainable Development (2005–2014)." UNESCO, Paris, n.d.

98 Victorian Association of Environmental Education (VAEE). Newsletter of Victorian Association for Environmental Education, 5/2005. Melbourne: VAEE, 2005.

99 Victorian Department of Education and Training (VDET). Commonwealth Games Education Project Plan. Melbourne: VDET,2004.

100 Victorian Government. Our Water Our Future: Securing our Waters Future. Melbourne: Victorian Government, n.d. Accessed March 21, 2007. http://www.ourwater.vic.gov.au/ ourwater/ about_our_water.

101 Victorian Parliament. Commonwealth Games Arrangements Act No. 57, 2001. Melbourne: Victorian Parliament, 2001.

102 VLAA. "Home." VLAA, Melbourne, n.d. Accessed December 15, 2009. http://www.litter.vic.gov.au/ /html/20-home-page.asp.

103 VLAA. "A Litter-Free Commonwealth Games." VLAA, Melbourne, 2006. http://www.vic.gov.au/www/html/891-a-litter-free-commonwealth-games.asp.

104 Waterwatch Australia. "About Waterwatch." Waterwatch Australia, Sydney, n.d. Accessed February 29, 2008, 2007.http://www.waterwatch.nsw.gov.au/01_about/index.html.

105 WWF. Environmental Assessment of the Athens 2004 Olympic Games. London: WWF, 2004.http:// www.panda.org/ downloads/-general/olympicsscorecardenglish.doc.

106 WWF. WWF Works with Olympic Organizers to Ensure "Green" Games in Beijing. London: WWF, n.d. Accessed April 12, 2008. http://wwf.panda.org/?116040/WWF-works-with-Olympic- organizers-to-ensure-quotgreenquot-games-in-Beijing.

107 Yea Wetlands Committee of Management. Yea Wetlands Walk(Pamphlet). Yea, Victoria: Yea Wetlands Committee of Management, n.d. http://www.yea.com.au/tourism/ Wetlands%20Pamphlet.pdf.

CHAPTER 6

Career fairs: are they valuable events? Hospitality and tourism recruiter perceptions of attributes towards participation and activities

Susan Gordon[1], Howard Adler[1], Sheila Scott-Halsell[2]

[1]School of Hospitality and Tourism Management,

Purdue University, Marriott Hall, West Lafayette, Indiana 47907, USA

[2] School of Hotel and Restaurant Administration,

Oklahoma State University, 210 Human Sciences Building, Stillwater, Oklahoma 74078, USA

ABSTRACT

The study examined the experience of hospitality and tourism industry recruiters who attended annual university career fairs sponsored by hotel and tourism departments at two major universities. The purpose of the study was to ascertain recruiter perceptions of the value of attending career fairs as a component of their companies' recruiting strategy. Recruiters attending the annual career fairs completed questionnaires. Survey results showed that recruiters perceive that career fairs are important to their companies' recruiting strategies and therefore, a valuable use of their time. Company representatives indicated that the universities were doing a good job in most aspects of the career fair with the exception of having students better prepared for the companies participating.

KEYWORDS

campus recruiters, hospitality, recruiter perceptions, career fairs, tourism

1 INTRODUCTION

Career fairs held on university campuses are one way for companies to meet recruiting goals, especially given the competition to attract the best talent (Silkes et al., 2010). Career fairs also allow students to learn about organisations and their recruiting needs (Brennan et al., 2004). There has been research about recruiters' behaviours and their effect on students (Wildes and Tepeci 2003), but no studies exist about recruiters' perceptions of the value in attending university career fairs. According to Chi and Gursoy (2009), the success of career services depends on the relationship that universities have with industry partners and, therefore, "it is important to explore factors that are likely to influence the success of the career services provided by hospitality programs" (p.309). Career fairs are a component of university career services, but since no research currently exists concerning recruiters' perceptions of career fairs, the present study is relevant to the success of university career services. Companies spend large amounts of money on recruiting, of which career fairs are a component, in order to secure future employees (Tepeci and Wildes, 2002). If recruiters do not feel that their time is well spent at career fairs, they may choose not to attend them. There are over 200 hospitality and tourism programs at universities in the USA (Crawford et al., 2009) so part of determining from where to recruit is based on university career fairs. Because of this, universities must provide effective career fairs as they cannot rely solely on their hospitality department's reputation to attract companies (Chi and Gursoy, 2009).

The purpose of the present research was to augment research about university recruiting and provide information to university career services on how recruiters from the hospitality and tourism industry perceive university career fairs. Specifically, the study attempted to reveal aspects of career fairs that recruiters found useful as well as any changes in career fairs that recruiters determined would strengthen the career fair experience. It is expected that universities will use the findings to make improvements to career fairs so companies will continue to attend them to recruit high quality candidates for positions. The study looked at the experiences of recruiters from hospitality companies who attended career fairs held by two major hospitality and tourism programs. The following research questions were answered in the study.

- What is the demographic profile of individuals who recruit at hospitality career fairs?
- Do recruiters feel that their attendance at career fairs is beneficial to their companies?
- Do recruiters feel that career fair facilities serve their requirements?
- Do recruiters feel that the students are well prepared?

- What aspects of career fairs would recruiters change?

2 LITERATURE REVIEW

2.1 Company recruiting strategy

Companies use career fairs as a means to find candidates with a balance of work and academic experience while also reducing the time it takes to fill positions (Hansen, 2006). In turn, career fairs allow students to learn about different companies and assess their fit with the organisation (Silkes et al., 2010). Another advantage of career fairs is that they allow students to learn about careers and career paths within an industry that they may not have realised existed prior to the fair (Payne and Sumter, 2005). Through these fairs, companies can establish contact with students and fill internships in addition to hiring graduating students for permanent positions (Hansen, 2006). Students benefit from a large number of companies gathering in the same location and companies benefit because they are able to access a large number of students during one visit (Payne and Sumter, 2005).

Recruitment of college students typically is conducted in cycles, which means that it is easier for a company to employ a planned strategy for this type of recruiting (Rynes and Boudreau, 1986). One study found that 66% of companies reevaluate college visit selections annually, which results in about 14% of companies adding campuses and 20% of companies dropping campuses (Rynes and Boudreau, 1986). The study found that the primary reasons for dropping campuses were lack of success in finding candidates, low job acceptance rates, and reductions in company recruiting expenses (Rynes and Boudreau, 1986). This may be especially crucial information for universities given the current economy and the fluctuation in college visits by companies. For many organisations, college recruiting can represent a large portion of their human resources budgets. The estimated average cost was 16% of the human resource budgets for Fortune 500 companies (Stevens, 1997). Most of the money was spent on recruiters' salaries and travel expenses (Stevens, 1997). In addition, if the company selects the wrong candidate into an entry-level management position, turnover can cost the company $50,000 or more (Tepeci and Wildes, 2002). Therefore, spending money to attend a career fair to find the 'best fit' candidates can be money well spent. One recruiter stated that spending money on career fairs overall is well worth the effort given the amount of exposure a company can obtain and the number of people it can reach (Hansen, 2006).

Companies also utilise online job fairs as a way to reach candidates and some smaller universities also utilise this format (Zelenskaya and Singh, 2011). Some companies have taken the online job fairs one step further by employing virtual programs that allow the recruiter and applicant to interact and conduct interviews in an online world (Zelenskaya and Singh, 2011). Because many companies do recruit online, companies have found that the combination of fairs and technology strengthens their recruiting strategies as fairs allow them 'face' time with quality candidates (Hansen, 2006). College recruitment involves a presence on college campuses including sponsoring events, giving company presentations, speaking at student organisations, and attending career fairs. A presence on campus is a tool to attract entry-level managers (Tepeci and Wildes, 2002). An on-campus interview is typically the first face-to-face contact between a recruiter and a student (Powell and Goulet, 1996). Interacting with a recruiter at a career fair is a way for students to meet the recruiter prior to the interview and, hopefully, make a more meaningful impression. However, companies must be careful not to become overexposed on campus. One study found that on-campus presence did not correlate with intention or behaviour to pursue positions at the company, suggesting that overexposure can cause perceptions of distrust and lowered credibility (Jaidi et al., 2011).

2.2 Recruitment of hospitality students

Throughout the last two decades, companies have implemented extensive on-campus recruiting processes to fill a growing skills shortage with qualified candidates (Miranda, 1999). Some companies may be on-campus year round as part of their corporate recruiting strategies (Chi and Gursoy, 2009). Hospitality schools have seen an increase in the number of companies attending their career fairs (Crawford et al., 2009). In addition, as the number of hospitality programs continues to increase (currently over 200 within the USA), the candidate pool will increase (Crawford et al., 2009). Hospitality recruiters are looking for candidates who have both industry skills and academic qualifications (Chi and Gursoy, 2009). In a recent study of hospitality recruiters of college graduates, the top five factors that recruiters in the hospitality industry are seeking when they visit a college campus were determined to be: "(1) how students prepare themselves for the real world (via internships), (2) level of faculty industry experience, (3) the quality of students' preparation for interviews for full-time jobs and internships, (4) reputation of the hospitality program, and (5) the quality of students' educational curriculum and courses taught in hospitality/tourism management" [Chi and Gursoy, (2009), p.314].

Some of the factors that influence hiring of hospitality graduates include: "student characteristics, recruiting-selection process, recruiters' demographic backgrounds, and their expectations" [Kwok et al., (2011), p.375]. The main criteria used by recruiters are student characteristics encompassing intellectual abilities, grade point average or academic accomplishments, work and internship experience, and activities outside of school such as clubs and sports

(Kwok et al., 2011). In a study by Kwok et al. (2011) of hospitality recruiters, other factors found to influence the hiring of hospitality students were personality, flexibility, leadership skills, and soft skills. When asked to rank the items most important in recruitment of hospitality students, leadership, relevant job experience, and person-organisation and person-job fit tied for the top item with personality ranked second (Kwok et al., 2011).

2.3 Recruiter profile

The population and experience of college recruiters varies and companies utilise different strategies to select who will visit college campuses (Harris et al., 2006). Recruiters may be recent college graduates who may relate better to students, seasoned managers to whom the new hires will report, or human resource directors (Harris et al., 2006). One study found that criteria for recruiters that were rated as important included: "(1) strong interpersonal skills, (2) enthusiasm for the company, (3) knowledge of company and jobs, and (4) credibility with students and co-workers" [Rynes and Boudreau, (1986), pp.740–741]. While the criteria of the recruiter seem to be important to companies, training does not draw the same interest. That study also found that only 41% of the companies surveyed offered standardised training programs, and only 48% required the training of recruiters (Rynes and Boudreau, 1986).

Companies send between one and five people to career fairs for an average of two days. These recruiters typically either work in human resources or operations (Silkes et al., 2010). The position that the recruiter holds within the company varies from company to company, but one effective method is to use employees within a university's geographic region as these employees can conduct on-site interviews at the school as well as establish ongoing relationships (Morley, 2001). Companies also send recent graduates to career fairs at the universities they attended because these employees already have connections with current students which may lead to a connection for the whole company (Hansen, 2006). Some companies have found that paying a stipend to a former intern to become a campus representative upon his return to school is also an effective way to increase awareness of the company (Hansen, 2006). Sending employees to career fairs not only increases employee morale, but it can also turn employees into year-round recruiters (Hansen, 2006). Another advantage of career fairs is the ability of employers to observe the competition, especially when attending industry-specific fairs.

2.4 Student perceptions of recruiters and companies

Career fairs provide recruiting opportunities, but they also allow companies an opportunity to increase brand awareness (Hansen, 2006). This is an important opportunity because some students only know about the company through the recruiter (Silkes et al., 2010). Applicants are influenced by "(1) how amiable, empathetic, caring, concerned, and how knowledgeable a recruiter is about the company, (2) its positions, and (3) its applicants" [Silkes et al., (2010)0, p.121]. In addition, research has shown that the impressions applicants have of the recruiters affect how applicants approach the recruiting process, including their decision about job selection (Connerley and Rynes, 1997). One study found that likability of hosts from the company that assisted students during site visits to potential employers explained the variance in job offer decisions (Turban et al., 1995). How the candidate feels about the host and those companies reflects that company representatives need to be selected carefully; (Turban et al., 1995), the same care needs to be taken of recruiters attending a career fair.

Signalling theory is one explanation of applicant job decisions since experiences with recruiters can serve as major, but unobservable, signals of organisational characteristics (Connerley and Rynes, 1997). For example, "interviewees interpret recruiters' interpersonal effectiveness (e.g., non-verbal cues, symbols, friendliness, organisational skills, and verbal behaviour) as signals of organisational effectiveness" [Wildes and Tepeci, (2003), p.41]. Prior studies have shown that the amount of information provided to candidates positively affects the attractiveness of the company as a place to work (Turban et al., 1995). In addition, students may perceive the absence of information as a signal that the company does not care enough to provide them with accurate information. Because research has shown that there is a strong relationship between applicant impressions during the early stages of recruiting and final decisions, impressions that recruiters make on applicants are of key importance (Connerley and Rynes, 1997). One major study found that one third of the students interviewed stated that the main reason they accepted the job offer was because of their interaction with the recruiter (Rynes and Miller, 1983).

2.5 Recruiter-educator relationships

Work expectations are ever-changing and recruiters are the connection between education and industry (Miranda, 1999). One of the ways students benefit from the relationship between recruiters and educators is that recruiters can give advice about which courses students should take to better prepare themselves for the area in which they are specialising (Payne and Sumter, 2005). Educators also need to keep up with industry changes in order to ensure that students are well prepared for future positions (Miranda, 1999). One way this is accomplished is through career fairs and relationships that are built between universities and recruiters. Often department-run career fairs are organised by students, allowing them to better learn about the industry they will be entering (Payne and Sumter, 2005). Students should work with professors to reach out to their industry contacts to produce a quality career fair. In addition, internship and job opportunities for students strengthen the recruiter-educator relationship. It often leads to other partnerships such as research as well as increasing both the school's profile and company's profile in the minds of students (Chi and Gursoy, 2009). These opportunities lead to long-term relationships that will continue to optimise benefits for students in the future.

3 METHODOLOGY

3.1 Instrument design

After a thorough review of the literature, no other study had been conducted about recruiter perspectives of career fairs. A focus group was held with recruiters at a large Midwestern university regarding issues of concern about career fairs. Results of the focus group and related industry literature review were used to develop a survey with questions about recruiter experiences at career fairs. In addition, various demographic questions were posed. The survey was divided into three sections and used various scales to measure responses.

The first section of the survey asked recruiters questions about their basic information, which included:

1 whether they were a graduate of a hospitality and tourism management (HTM) program

2 their title

3 number of times they had attended the particular career fair at which they received the survey

4 average number of days they spend per year at career fairs

5 average number of career fairs attended

6 number of people sent to a career fair by their company

7 if interviews are conducted while at the career fair

8 are the interviews for internships, graduating students, or both.

This section also asked recruiters about their companies' recent hiring of college graduates in terms of how it has changed during the previous year. Recruiters were asked if they were looking for the same or different number of students and if the number of university career fairs they attended had changed.

The second section of the survey asked about the value of career fairs as a recruiting tool for their companies and their opinions on the format of career fairs. Recruiters were asked to rate the value that they placed on attending hospitality career fairs as well as the main purpose of attending them. The questions were developed from the 2006 study conducted by Harris et al.

In the third section, recruiters were asked what they liked best about career fairs related to such activities as:

1 keynote speakers

2 evening receptions

3 lunch with students

4 interviewing while at the fair

5 interacting with other recruiters

6 interacting with students.

These questions were developed based on the 2009 article by Harrison et al. The final question was designed to solicit possible changes recruiters may suggest regarding career fairs including:

1 length of time

2 attendance regulated by student class (freshman, sophomore, junior, and senior)

3 size of booth

4 number of students in attendance.

3.2 Data collection and analysis

The present study focused on recruiters who attended annual HTM program career fairs held at two large Mid western universities. Recruiters who attended the two career fairs were surveyed regarding their perceptions of career fairs in general and the effectiveness of career fairs as a tool to meet their companies' recruiting needs. Those recruiters primarily were human resource and operations professionals from a total of 68 different hospitality companies. Among the companies represented at the fairs, 32 were classified in the lodging sector, 29 were classified as foodservice or restaurant operations, four were classified as tourism related, two companies were in the foodservice equipment sector, and the last one was in hospital management. Many of the companies had more than one person from their organisation attending. The population of recruiters for the two career fairs was 125. The research was conducted during the fall of 2010 and spring of 2011. These particular career fairs are held once per year as opposed to some universities which hold career fairs during both semesters. One university held its career fair in the fall and the other one held its event during the spring for three hours in the afternoon. All students within the hospitality and tourism departments were invited to attend. A reception for the recruiters was held the evening prior to each career fair with the hospitality faculty and graduating seniors invited to meet recruiters.

The survey was distributed to recruiters during the reception the evening prior to the career fair. The survey included a letter that advised the participants of the purpose of the research and assured them of confidentiality and anonymity of their responses. The letter also provided participants with the researchers' names and contact information. Those recruiters who did not attend the reception were given the letter and survey the following day during the career fair. Completed surveys were collected during the fair; those recruiters who were unable to complete the survey during the career fair returned their responses by postage paid envelopes. The research was conducted independently from any participating companies, although results are available by request from participating companies. Survey data was entered into an Excel database for quantitative analysis.

4 RESULTS AND DISCUSSION

Findings of the survey related to recruiter perceptions and answered the research objectives. Do recruiters feel that career fairs are valuable, what are the most important activities they value as part of the career fair such as networking, and interviewing, and do HTM programs facilitate these activities? If they do not find the career fair valuable, what can HTM programs do to change that perception as they should not rely simply on the department or university's reputation to attract recruiters (Chi and Gursoy, 2009). Such information will allow HTM programs to change the format or activities of career fairs to better accommodate recruiter attendees in the future.

4.1 Recruiter and company demographics

A total of 125 surveys were distributed to recruiters; 76 were returned for a response rate of 61%. Of those who responded, 50% were male, 34% were female, and 16% did not provide their gender. The majority of respondents were from hotel companies at 49% of the total, followed by restaurants at 38%, and others including healthcare, suppliers, convention bureaus, and retail at 13%. In terms of the positions held by recruiters, human resource individuals accounted for 57% of the respondents, operations accounted for 39%, and 4% of respondents did not give their job title. Only 42% of the respondents were graduates of a hospitality program, while 58% had other majors.

Recruiters were asked how many times they had attended the particular career fair, including the current visit. Of the 76 respondents, 33% were attending the fair for the first time, 17% for the second time, 20% for the third time, 27% for the fourth time or more, and 3% did not respond to the question. Recruiters also were asked the average number of days they spend at a university when attending a career fair. The majority (61%) indicated they spend two to three days, 36% indicated one day, and 3% did not respond to the question. When asked the average number of people sent to the career fair by the companies, 11% indicated one person, 43% indicated two people, 33% indicated three people, 11% indicated four people, and 2% did not respond. Recruiters were then asked how many career fairs at other universities they attend per year. Of the 76 respondents, 20% responded that this was their only career fair, 14% attended one other career fair, 24% attended two or three other career fairs, 24% attended between four and nine other career fairs, 14% attended ten or more other career fairs, and 4% did not respond to the question.

Almost three-fourths (71%) of the respondents indicated they were also conducting interviews while they were on campus. Of the 54 respondents who were interviewing, nine were interviewing only for internships and no May graduates, three were allocating 10% or less of their interviews for May graduates and the rest were for internships, four were allocating 20% to 30% for May graduates, 19 were splitting the interviews at 50% for May graduates and 50% for internships, 13 were allocating 60% to 95% for May graduates, four were dedicating all interviews for May graduates, and two recruiters did not respond.

In terms of their recruiting efforts during the last year, 47% of the respondents reported their company was visiting the same number of schools as the previous year, while 36% reported their company was visiting more schools, and 17% reported that their company was visiting fewer schools. Compared to the previous year, 31% of the recruiters indicated that their company was hiring more graduates, 47% stated their company was hiring the same number of graduates, and 22% stated their company was hiring fewer graduates.

4.2 Career fair value

Regarding value, recruiters were asked if they felt that career fairs are an important component of their company's recruiting strategy.

The mean response was 4.5 on a scale of 1 to 5, with the higher number indicating very important. Recruiters believed that attending career fairs on behalf of their company is an important function. Recruiters also were asked if they felt that career fairs are mainly a public relations (PR) function for a company. Recruiters averaged a 3.07 response on a scale of 1 to 5, indicating that some feel that it is simply a PR function to keep a company's name relevant while some do not hold that belief. Possibly while some recruiters agreed that career fairs only serve a PR function, they did not see this as a negative aspect in terms of recruiting, especially given that many of the companies are not currently in a position to hire numerous people. When asked if recruiters felt that career fairs are a good use of company time and a way to make the company better known, the mean response was 4.39 on a scale of 1 to 5. When asked about student preparedness, recruiters felt that students were well prepared to represent themselves to companies as indicated by the mean response of 4.13 on a scale of 1 to 5. Table 1 addresses recruiter perceptions of career fair value.

Table 1 Recruiter perceptions of career fair value

Perceptions of career fair value	
	Mean response
Important component to recruiting strategy	4.50
Help make company become better known	4.39
Happy with career fair format	4.28
Student were prepared	4.13
Public relations activity for company	3.07

4.3 Career fair environment

The researchers asked about career fair facilities. The majority of recruiters felt that the facilities of the career fairs met their needs. The mean response was 4.38 on a scale of 1 to 5, with 1 as strongly disagree and 5 as strongly agree. Regarding the format of career fairs, the mean response was 4.28 on a scale of 1 to 5, indicating that overall recruiters were satisfied with how the two schools run their career fairs. Only 36% of the recruiters indicated that they would change some aspect of the career fairs. Among the 27 who indicated they would make changes, 17 selected regulating attendance by class (freshman, sophomore, junior, and senior), as the most popular answer followed by a desire to have a longer career fair as indicated by 15%. A few respondents wrote in their own suggestions. Some of the comments were "more time to interview, one hour not 30 minutes", "more attendance by students", "more interaction with university faculty", "each company [makes] a presentation and/or small group panel discussions", and "have [company]information session prior to career day". Table 2 shows recruiter preferences for changes to career fair formats.

Table 2 Recruiter preferences for changes to career fair formats

	Perceptions of format changes	
	Frequency n = 27	*Percent*
Regulate attendance	17	63%
Longer career fair	4	15%
Shorter career fair	3	11%
Larger booths	3	11%
Less crowded	0	0%

4.4 Career fair activities

In the third section of the survey, recruiters were asked to rank the top two activities they deemed most useful during the career fair. The majority selected interacting with students as their first choice and being able to interview students as their second choice. The responses demonstrated that recruiters' main objective in attending career fairs was to interact with the students, both at the career fair itself and while interviewing. Table 3 shows recruiter perceptions of career fair activities.

Table 3 Perceptions of career fair activities

	Perception of career fair activities			
	Frequency n = 66			
	First choice	*Percent*	*Second choice*	*Percent*
Interact with students	47	71%	12	18%
Time to interview	13	20%	30	45%
Keynote speakers	5	8%	3	5%
Evening receptions	1	2%	7	11%
Lunch with students	0	0%	7	11%
Interact with recruiters	0	0%	7	11%

5 CONCLUSIONS AND RECOMMENDATIONS

Many university hospitality programs have their own career centres to aid students in securing internships and jobs upon graduation (Chi and Gursoy, 2009). Career fairs typically are a component of these career services. Career fairs are a tool commonly employed by various university departments to showcase their students. At the same time, companies are able to go to the career fair and use the advantage of a large number of students to meet their future workforce needs. Because recruiters are vital to the success of student recruiting and career fairs, understanding recruiter perceptions of career fairs is important in order to ensure they attend them. If companies send recruiters to career fairs, then students may perceive that the company places value on the career fair and the students who attend them (Rynes et al., 1980). If companies understand that students have a better perception of them due to their career fair participation and attend career fairs due to company participation, then companies will be more apt to continue to attend career fairs. The perceptions of recruiters about career fairs is also crucial to the administration of hospitality departments as placement rates, location, and position can help attract future applicants to the school (Kwok et al., 2011). In terms of academic contributions, while research has been conducted about students' perceptions of career fairs (Silkes et al., 2010), to the authors' knowledge, recruiters' perceptions of career fairs have not been studied previously. Therefore, the present study adds to research about university recruiting by evaluating the perceptions of recruiters at career fairs.

One of the recruiting strategies that companies utilise in college recruiting is to send operations managers to career fairs within their geographic area (Morley, 2001). Another interesting finding from the career fair survey was that of the 26 recruiters who do not attend another career fair or only attend one other career fair, only four of those respondents were in human resources. By comparison, of the 29 respondents who attend four or more other career fairs, only two of them were from operations. Such findings would suggest that hospitality and tourism companies are utilising the strategy of sending operations managers to some career fairs within their local area, while others send human resource managers to several career fairs regardless of location. This result is also consistent with the findings in the previous study (Silkes et al., 2010) that recruiters typically either work in human resources or operations.

The study found that 87% of the respondents indicated that their company sends more than one person to a university career fair. In addition, 61% of the respondents were spending two or three days on campus to attend the career fairs and recruits either before or afterwards. The findings were also evident in the previous literature (e.g., Silkes et al., 2010). The present study suggests that universities need to be aware of the number of days recruiters are spending on campus and take advantage of it. For example, more opportunities for partnerships with faculty that go beyond speaking in the classrooms can be developed, such as consulting work that may also produce increased research.

Since recruiters indicated what they enjoyed most about career fairs was interacting with students, hospitality and tourism programs could develop additional opportunities such as forums or panels of recruiters. Commonly practiced, recruiters appear in the classroom, but this often takes a more formal approach. In addition, students often voice the desire to have more informal interaction with company recruiters to learn more about the industry and the successful career paths of the recruiters themselves who often are engaged in higher-level management. This could be accomplished through inviting companies to sponsor school events or asking them to speak at student clubs, which also benefits companies since it allows them to establish an on-campus presence (Tepeci and Wildes, 2002). Only one respondent indicated that interacting with other recruiters was an aspect he or she liked best; recruiters are more focused on students than they are with networking among their peers during career fairs. Therefore, universities would do well to increase the opportunities for student-recruiter and faculty-recruiter interaction rather than trying to increase networking for recruiters.

Another comment related to the time spent on interviews. Recruiters would like to extend the timeframe for each interview. Both students and recruiters would benefit; however it would be necessary for companies to extend the amount of time that they spend at each school or send more recruiters. This is often not feasible due to budget and time constraints as travel expenses comprise a large amount of companies' human resource budget (Stevens, 1997).

While career fairs are an important part of many companies' recruiting efforts, tight human resource budgets often call for companies to be very selective in the number of schools they visit. A reduction in recruiting expenditures is one of the reasons companies decide not to continue visiting a campus (Rynes and Boudreau, 1986). Another reason for no longer going to a university is lack of ability to find the right candidates (Rynes and Boudreau, 1986). Some of the recruiters' comments pertaining to the students were "students should research [the] company before the fair", "higher student attendance", and "survey students as to what they are looking for in a company and why they select one over another". Such comments suggest that universities could do a better job of encouraging students to not only attend the fair, but to research the companies before speaking with recruiters. Because industry experience is important to hospitality and tourism industry recruiters, educators must continue to ensure that students are prepared for the career fair to secure internships and permanent jobs. In addition, the students' preparation reflects on the university; so students must present themselves well when interacting with recruiters.

6 LIMITATIONS AND RECOMMENDATIONS FOR FUTURE RESEARCH

The study represented only two career fairs at two universities. While the study was representative of the recruiters who attended the career fairs at the universities where the study was conducted, the possibility exists for differences with other similar programs and fairs based on the type of department in which the program is housed, the university's size, location of the university, and others. However, both of the programs have been in existence for over 50 years, are ranked in the top 10 among hospitality programs (Severt et al., 2009), and hold career fairs which are heavily attended by major companies in the hospitality industry. Therefore, these programs are viewed as highly generalisable within the hospitality program university setting. One of the career fairs was held in the fall and the other university held its event in the spring, so there may be additional differences based on this fact. In the fall, sometimes recruiters do not interview during the career fair and choose to return to campus during the spring to interview. Also during the spring, students who have not yet found employment may tend to be more aggressive with the time that they spend with recruiters. These factors may affect recruiter attitudes. Future research could study the differences in recruiters' perceptions of career fairs based on time of year held or differences in recruiting efforts in terms of types of interviews conducted (internships or May graduates). In addition, future research could focus on the success that companies have had with the students they recruited through career fairs over time.

AUTHOR DETAILS

Susan Gordon is a PhD student and graduate research assistant at the School of Hospitality and Tourism Management, Purdue University. She received her Masters in Business Administration from Boston University and also holds a Bachelor of Science in Hospitality Administration from the University of Massachusetts at Amherst. Her industry work experience spans over ten years and includes management positions with Starwood Hotels and Resorts, Marriott International, Inc., and Walt Disney World. Her research interests include human resources and lodging operations.

Howard Adler has been a faculty member at Purdue University since 1982 teaching primarily in the area of hotel operations. He has held several management positions in the hotel industry including General Manager with Sheraton as well as Director of Sales, Front Office Manager and Resident Manager with some private resorts. He spent the 1999/2000 school year teaching at the University Del Este, in San Juan, Puerto Rico and the 2006/2007 school year teaching at the South China University of Technology in Guangzhou, China and the Macau Institute of Tourism, Macau, China.

Sheila Scott-Halsell is an Associate Professor and Associate Director of the School of Hotel and Restaurant Administration at Oklahoma State University. She is also the co-Director of the US-EU FIPSE HOTEL Atlantis Dual Degree Program. Before joining OSU in 2006, she received her PhD from Texas Tech University. Spending 17 years in hospitality management roles, prior to returning for graduate studies, strongly influences her research areas of human resources, organisational behaviour and event management. Her grant activities, as PI or co-PI, total $501,607.23.

REFERENCES

1 Brennan, C. , Daly, M. , Fitzpatrick, E. , Sweeney, E. (2004). 'Tradition vs. technology: career fairs in the 21st century'. International Journal for Educational and Vocational Guidance. 4, 1, 23-41

2 Chi, C. , Gursoy, D. (2009). 'How to help your graduates secure better jobs? An industry perspective'.International Journal of Contemporary Hospitality Management. 21, 3, 308-322

3 Connerley, M. , Rynes, S. (1997). 'The influence of recruiter characteristics and organizational recruitment support on perceived recruiter effectiveness: views from applicants and recruiters'. Human Relations. 50, 12,1563-1586

4 Crawford, A. , Hubbard, S. , Gaillard, N. , Waln, M. (2009). 'Strategies for on-campus recruitment: a decade later'. Consortium Journal of Hospitality & Tourism. 13, 2, 5-13

5 Hansen, F. (2006). 'Far from obsolete, career fairs are a low-cost way for firms to hire quickly and fill pipelines'. Workforce Management. 85, 4, 46-47

6 Harris, K. , Kwansa, F. , Lattuca, F. (2006). 'Recruiter opinion of hospitality programs: an assessment of selected programs'. Journal of Human Resources in Hospitality & Tourism. 5, 1, 17-31

7 Harrison, K. , Kerby, D. , Fleak, S.K. (2009). 'Foraging effective campus relationships'. The CPA Journal. 79, 1, 69-71

8 Jaidi, Y. , Van Hooft, E. , Arends, L. (2011). 'Recruiting highly educated graduates: a study on the relationship between recruitment information sources, the theory of planned behavior, and actual job pursuit'. Human Performance. 24, 2, 135-157

9 Kwok, L. , Adams, C.R. , Price, M.A. (2011). 'Factors influencing hospitality recruiters' hiring decisions in college recruiting'. Journal Of Human Resources In Hospitality &Tourism. 10, 4, 372-399

10 Miranda, P. (1999). 'The Level of technical skills and management competency demanded by the hospitality industry as perceived by hospitality recruiters'. Menomonie, WI:University of Wisconsin-Stout , Unpublished master's thesis, retrieved September 6, 2010 from Dissertations and Theses database

11 Morley, M. (2001). 'Opening the window of job opportunity'. Lodging Hospitality. 57, 11, 28

12 Payne, B. , Sumter, M. (2005). 'College students' perceptions about career fairs: what they like, what they gain, and what they want to see'. College Student Journal. 39, 2, 269-276

13 Powell, G. , Goulet, L. (1996). 'Recruiters' and applicants' reactions to campus interviews and employment decisions'. Academy of Management Journal. 39, 6, 1619-1640

14 Rynes, S. , Boudreau, J. (1986). 'College recruiting in large organizations: practice, evaluation, and research implications'. Personnel Psychology. 39, 4, 729-757

15 Rynes, S. , Miller, H. (1983). 'Recruiter and job influences on candidates for employment'. Journal of Applied Psychology. 68, 1, 147-154

16 Rynes, S. , Heneman, H.G. , Schwab, D.P. (1980). 'Individual reactions to organization recruiting, a review'.Personnel Psychology. 33, 3, 529-542

17 Severt, D.E. , Tesone, D.V. , Bottorff, T.J. , Carpenter, M.L. (2009). 'A world ranking of the top 100 hospitality and tourism programs'. Journal of Hospitality & Tourism Research. 33, 4, 451-470

18 Silkes, C. , Adler, H. , Phillips, P.S. (2010). 'Hospitality career fairs: student perceptions of value and usefulness'. Journal of Human Resources in Hospitality & Tourism. 9, 2, 117-130

19 Stevens, C. (1997). 'Effects of preinterview beliefs on applicants' reactions to campus interviews'. Academy of Management Journal. 40, 4, 947-966

20 Tepeci, M. , Wildes, V. (2002). 'Recruiting the best: a study in attracting hospitality management students for entry-level management positions'. Journal of Quality Assurance in Hospitality & Tourism. 3, 1–2, 95-107

21 Turban, D. , Campion, J. , Eyring, A. (1995). 'Factors related to job acceptance decisions of college recruits'.Journal of Vocational Behavior. 47, 2, 193-213

22 Wildes, V. , Tepeci, M. (2003). 'Influences of campus recruiting on applicant attraction to hospitality companies'. Journal of Human Resources in Hospitality & Tourism. 2, 1, 39-51

CHAPTER 7

The volunteer legacy of a major sport event

Alison Doherty

Faculty of Health Sciences, School of Kinesiology , University of Western Ontario , Canada

ABSTRACT

The purpose of this study was to understand the volunteer legacy of a major sport event and identify aspects of the event that shaped future voluntary action in the host community. Social exchange theory framed the examination of volunteers' positive and negative experiences with the event as a predictor of future behavioral intentions. A total of 1098 volunteers involved with the 2001 Canada Summer Games completed a post-event survey. In general, planning volunteers' future volunteering was particularly influenced by experienced costs of the event (task overload, personal inconvenience), although contributing to the community and a positive life experience were also predictive of their future involvement. In contrast, on-site volunteers' future volunteering was more influenced by experienced benefits of the event, including social enrichment, community contribution, and a positive life experience. However, personal inconvenience and task underload were also predictive of their future involvement. The findings have implications for event policy and management that should acknowledge the potential for major sport events to engender a legacy of volunteering.

KEYWORDS

event management, community development, volunteerism

1. INTRODUCTION

Major sport events can bring vitality to a community. There may be considerable economic benefits from visitor spending, increased community profile, and job creation (Gratton, Shibli, & Coleman, 2005; Gratton & Taylor, 2000; Misener & Mason,2006). A community may also benefit from capital development where there are new and upgraded athletic facilities, as well as any improvements to roads or parks as part of hosting an event. Gratton and Taylor (2000) also noted the social benefits to a community of such capital development associated with an event. They further suggested that social benefits, or what Crompton (2004) referred to as 'psychic income,' may be accrued through improved community self-image. Major sport events can provide a venue for the expression of community pride, and can be a rallying point for the community to celebrate. Given these potential benefits, 'many [municipal leaders] have subscribed to the notion that sport can be a catalyst for urban growth and have employed various scales and sizes of sporting events to showcase their respective cities to investors, tourists, and residents alike' (Misener & Mason, 2008, p. 604).

Major sport events refer to one-time or perhaps annual sport competitions, as opposed to leagues. A key difference is that an event is of shorter duration (perhaps a week, a weekend, or even a day), and tends to be higher profile than a sports league; consider, for example, the profile of the National Football League and then its premiere event, the Super Bowl Game. The size, focus, and profile of major sport events vary; however, an increasing number of such events, and their potential role in community development, demands consideration of their management and impact in the community. While major sport events have a fundamental focus on the competition itself, the legacy of such events is increasingly recognized as having the potential to extend beyond this objective (e.g., Downward & Ralston, 2006; Hiller, 2006; Matheson & Baade, 2006; Sterken,2006). The capital, economic, and social benefits that can be realized from event hosting have elevated national, regional, and community interest in major sport events (Gratton et al., 2005; Misener & Mason, 2006, 2008).

These major sport events, and their payoffs, would not be possible without the efforts of community volunteers (Cuskelly, Hoye, & Auld, 2006). In turn, major event hosting provides an opportunity for individual and community development through volunteer work (Downward & Ralston, 2006). Indeed, community development may be a formal objective of a host committee, in terms of leaving a legacy of volunteerism in a community. Such a legacy may be defined as the carryover effect of ongoing community support (Lynch, 2001). It may be indicated by increased volunteer support for other special events in the community and enhanced volunteerism in the community in general, in terms of positive attitudes towards volunteering, an increased rate of volunteering, and increased level of involvement. But, what impact does a major sport event have on future volunteerism? And what aspects of the event contribute to, or detract from, that legacy? Whether or not a volunteer legacy is a formal objective of hosting an event, it is important to understand what about the volunteers' experience might influence their future volunteering, with implications for event policy and management.

Volunteer management is only one part of major sport event management, and the focus tends to be on determining at the outset where volunteers are needed and how many are required, then recruiting, training, and managing those volunteers to ensure the event is a success (e.g., Gladden, McDonald, & Barr, 2005). However, the impact of the event on future volunteering should be an important consideration. For example, major sport events may be an opportunity for community members to get involved in formal volunteering for the first time. Their experience there may translate into further community volunteering, either at another event or on a more regular basis with one or more community organizations. Future community events of any kind will only benefit from the involvement of volunteers who have previous experience with hosting and staging an event, and who may be willing to get involved to a greater extent (cf. Downward & Ralston, 2006). Given the ever increasing dependence of our communities on volunteerism (e.g., Brennan, 2005), it is important to identify any and all factors that contribute to participation.

There has been considerable research on what shapes voluntary action, including in the domain of major festivals and events (Cuskelly et al., 2006). However, little is known about what influence those events have on future voluntary action, if any (Cuskelly et al., 2006). Intent to engage in behavior is the best predictor of actual behavior (e.g., Sheeran, 2002; Tett & Meyer,1993); thus, it is useful to consider future intentions as an indicator of volunteer legacy. It is furthermore important to understand the factors that contribute to those behavioral intentions. The purpose of this study was to determine the influence of volunteers' experiences with a major sport event on their future volunteering intentions, expressed by whether they would volunteer for another major event, volunteer to a greater or lesser extent in a future event, and volunteer to a greater or lesser extent in the community in general. The study was framed by social exchange theory (Homans, 1958; Thibault & Kelley, 1959) which purports that perceived costs and benefits associated with a particular behavior predict intent to engage, or remain engaged, in that behavior.

There are two distinct groups of volunteers who work to make a major sport event happen: planning volunteers are involved from several months up to several years leading up to the event; on-site volunteers are involved during the event itself. Given their very different levels of involvement, a sub-purpose of this study was to determine whether there was any variation in the experiences and future intentions of planning volunteers and on-site volunteers. Also, given that previous involvement in volunteer activities is an important determinant of future volunteering (e.g., Hall, Lasby, Gumulka, & Tryon, 2006), another sub-purpose of the study was to determine whether there was any variation in the experiences and future intentions of individuals who had previous formal volunteering experience and those who were volunteering for the first time in a formal capacity. The paper now turns to a review of volunteer contributions to the successful staging of major sport events, a discussion of the contributions of event volunteering to community development, and a review of the extant research on event volunteers' experiences and intentions.

1.1. The contribution of major sport event volunteers

Both the 2000 Sydney and 2004 Athens Olympic Games relied on over 40,000 volunteers (Cuskelly et al., 2006), while the 2008 Beijing

Olympic Games had 100,000 volunteers involved with competition venues, the Olympic villages, and transportation (Yan & Chen, 2008). The 2010 Vancouver Olympics is expected to require up to 25,000 volunteers to help organize and stage the Games (Volunteer BC, 2008). On a smaller scale, city marathons may rely on as few as 500 and as many as 1000 volunteers to ensure their success. The economic contribution of these event volunteers can be considered. For example, the Sydney Olympics utilized 45,000 volunteers who contributed an average of 100 hours each (Lynch, 2001) or approximately 4.5 million hours. If those volunteers were paid at least a minimum wage of, say, $10/hour, the cost to the organizing committee for that volunteer labor would have been $45 million. Solberg (2003) conducted a more systematic analysis of the economic impact of event volunteers, focusing on the 1999 World Ice Hockey Championships. He determined that the 800 volunteers, who contributed more than 71,000 hours, provided a service with high market value, however their time spent volunteering did not compromise their regular contribution to the local economy.

It is also important to consider the non-economic or social impact of volunteers, taking into account the time, skills, enthusiasm, warmth, and passion of the hundreds or thousands of volunteers who become involved in an event of their own free will and for no financial remuneration (cf. Smith, 1981). Volunteers are often considered the 'face of the Games,' and organizers rely on a positive image that reflects well on the event and the organizing committee. That impact can also be realized in the larger community, as citizens become involved to support the event and make their community a better place. The dollar value of that may be inestimable.

1.2. Event volunteerism and community development

A legacy of community volunteerism is a worthwhile mandate of any host committee because, as Cnaan and Goldberg-Glen (1991, p. 270) noted, volunteers 'strengthen the demographic fabric of society.' Community development is 'a process of building relationships that increase the adaptive capacity of people who share a common locality' (Brennan, 2005, p. 20). This capacity reflects the ability of community members to 'voluntarily organize, manage, utilize, and enhance those resources available to them in addressing local needs' (Brennan, 2005, p. 20). Misener and Mason further suggested that this 'collective sense of responsibility [may be] generated by broad-based participation in community initiatives' (2006, p. 44). Major sport events bring large numbers of people together as volunteers. They create a sub-community of individuals connected through a collective mission to stage a successful community event. According to Brennan (2005), interaction among volunteers (such as with event volunteering), as they work towards and successfully contribute to common local needs, may be expected to engender further voluntary action to improve the social, cultural, and psychological circumstances in a community. Thus, volunteers' experiences with an event may be expected to shape future voluntary action.

1.3. Research on major sport event volunteers' experiences and future intentions

Several studies have examined volunteers' major sport event experiences, focusing particularly on their satisfaction with aspects of the event, or satisfaction overall. With a few exceptions, the research in this area has been descriptive. Elstad (1996) found that a sample of student volunteers involved with the 1994 Lillehammer Winter Olympics developed social skills, knowledge about society in general, and job-specific competencies as a result of their involvement. She reported high levels of satisfaction among the volunteers, with the most positive aspect being the opportunity to develop one's personal network through meeting new people and making friends; they were also particularly satisfied with the 'celebratory atmosphere.' The most commonly noted downsides to volunteering for the students were out of pocket expenses for transportation and food, having too little or too much to do, and poor organization of volunteers. Farrell, Johnston, and Twynam (1998) found that the majority of their sample of volunteers with the 1996 Canadian Women's Curling Championships were very satisfied with their experience overall. Communication with other volunteers and recognition for their contribution were significant predictors of that overall satisfaction. In a study of volunteers with the 2001 Francophone Games, Larocque, Gravelle, and Karlis (2002) reported a high level of satisfaction with their event experience overall. However, the volunteers were more satisfied with such aspects as quality of their volunteer team and recognition from the organizers, and relatively less satisfied with the quality and level of responsibilities they were assigned.

Building on the descriptive work to that point, Costa, Chalip, Green, and Simes (2006) studied volunteers with the Sunbelt IndyCarnival, an IndyCar race held annually on the Gold Coast, Australia, to identify antecedents of volunteer satisfaction with that event. Through path analysis, they found that the opportunity for volunteers to share their opinions and experiences during training contributed to their sense of community at the event itself, which positively impacted their commitment to the event. Further, higher levels of event commitment were associated with increased satisfaction with their volunteer role.

Only a few studies have examined future intentions to volunteer and, with the exception of Downward and Ralston (2006), this work has been purely descriptive. As part of their study of Francophone Games volunteers, Larocque et al. (2002) also reported that it was quite important to these individuals to maintain their involvement as volunteers. Twynam, Farrell, and Johnston (2002/2003) examined the future intentions of volunteers with the 1998 World Junior Curling Tournament and found that half (49%) indicated that the tournament experience would increase their likelihood of volunteering again, just less than half (45.5%) said their likelihood of volunteering would not change, and the remainder said their likelihood of volunteering would decrease. In a study of the 2005 Canadian Women's Open Golf Championships, MacLean and Hamm (2007) found that almost all of the volunteers intended to continue as golf volunteers (97.5%), 76.4% as sport volunteers, and 83.3% would continue volunteering in general. An open-ended question elicited various reasons for these behavioral intentions; however, no clear link was established between the volunteers' future intentions and their experiences with this event.

In an explanatory study, Downward and Ralston (2006) examined, in part, factors associated with 2002 Commonwealth Games volunteers' post-event volunteering intentions. They conducted a survey with a sample of volunteers about one year after the event and found that 85% and 68% of those individuals were willing to be involved with another major sports event and another major event in general, respectively, while 43% increased their interest in voluntary work in general following the Commonwealth Games. With respect to the impact of their event experiences, the survey asked the volunteers about the extent of (1) the organization of volunteers, (2) satisfaction with their task assignment, (3) their personal development, (4) their contribution to the community, and (5) their contact with celebrities. Downward and Ralston found that the personal development volunteers' experienced was a consistent predictor of their intention to volunteer for another event and to volunteer in general. There was no distinction in event experiences or their impact on future volunteering between those with previous volunteer experience and those with no previous experience.

1.4 Theoretical framework

The current study appears to be only the second to examine volunteers' experiences with a major sport event as a predictor of their future behavioral intentions. Such a focus will enhance our understanding of the legacy of major sport event volunteering and the factors that impact on that. To do so, the study relies on the basic tenets of social exchange theory (Homans, 1958; Thibault & Kelley, 1959), which argues that individuals enter into relationships with other individuals or institutions where they believe they can receive valued benefits, yet incur few costs. Benefits to the respective parties in the exchange may be tangible or extrinsic (e.g., economic, services, information) and intangible or intrinsic (e.g., social, emotional), thus taking social exchange theory beyond the concept of relationships that are based simply on economic or market transactions (Cropanzano & Mitchell, 2005). The basic assumption of exchange theory is that social relations are established and maintained based on the expectation that such relations will be mutually advantageous (Blau, 1994). As Zafirovski further noted, behavior is governed by reciprocal relations and 'would not continue in the long-run if reciprocity were violated' (2005, p. 3). Such violations would occur if the exchange ratio was imbalanced; that is, if an individual experienced fewer rewards than costs as a result of the relationship. As such, 'people participate in relationships and activities only so long as the outcomes of participation are sufficiently favourable' (Guillet, Sarrazin, Carpenter, Trouilloud, & Cury, 2002, p. 93).

Volunteering is inherently an exchange relationship, where individuals offer their time, skills, and energy to assist with an event, and experience various benefits, as well as costs, in return. As such, it may be assumed that major sport event volunteers will be more likely to engage in future volunteering behavior to the extent that they have experienced positive outcomes as a result of that behavior in the past; similarly, they will be less likely to volunteer again if they have experienced negative outcomes. A particular advantage of social exchange theory is that it facilitates a comprehensive understanding of the impact of both positive and negative experiences on (future) behavior (e.g., Guillet et al., 2002; Howard & Crompton, 2004; Waitt, 2003; Weiss & Stevens, 1993). The current study examined what positive and negative aspects of volunteers' experiences with one major sport event were most meaningful to their future intentions with regard to community volunteering. Consistent with volunteering as an unpaid activity, the focus of this study was socioemotional rather than economic benefits and costs (Cropanzano & Mitchell, 2005).

2. METHOD

2.1 Focal event

This study was the second phase of a project examining the pre- and post-event attitudes, behaviors, and experiences of volunteers with the 2001 Alliance London Jeux du Canada Games. The Canada Games is one of the largest youth sports events in the world, with 3000–4000 athletes competing in 15–20 sports. The typically junior-level athletes (average age of 16 years) qualify and compete for their provincial or territorial team, with all teams vying for the overall championship flag. The Games were first held in 1967 to celebrate Canada's Centennial. The Games are staged every two years and alternate between Summer and Winter Games. In 2001, the Canada Games was hosted by an alliance of communities in and around London, Ontario (population 340,000). The Games hosted over 5000 athletes and relied on a workforce of 6500 volunteers for the lead-up planning and on-site implementation of the event. A total of 22 venues were used for 17 different sports in these communities.

Like any major community festival or event, the success of this event was tied to the involvement of volunteers from the host community. Planning volunteers included the Games' Co-Chairs, and Vice Chairs only who were responsible for such functional areas as administration and finance, sport services, volunteer services, and the athletes' village. Associate Vice Chairs were responsible for various portfolios within each area and were assisted by sub-committees within each portfolio. Other planning volunteers included Sport Leaders and Venue Leaders who were responsible for the technical aspects of the sport competitions and logistics of the competition venues, respectively. Planning volunteers were involved for up to four years before the Games began. On-site volunteers were assigned to competition and administrative venues across the host communities. They were involved in everything from parking, to shuttling participants, from minor officiating, to hospitality during the Games.

The Host Society's vision was 'to advance the Canada Games' legacy of building Canadian unity, fostering personal excellence through sport and enriching community life' (2001 Alliance London Jeux du Canada Games, 2001, p. 4). Further, one of the stated missions of these Games was to achieve 'a legacy of enhanced sport development, bilingualism, facilities and volunteerism' (2001, p. 4). Thus, a planned legacy of the 2001 Canada Games was 'to create a strong volunteer base for the communities involved' (L. Foster, manager, volunteer services, personal communication, 10 July 2001). This focus is consistent with Canada's strong tradition of volunteerism, where 45% of the adult population give their time to formal organizations or groups, and 83% help others more informally (e.g., doing housework, driving to appointments) (Hall et al., 2006).

2.2 Participants

The total sample of 1098 volunteers for Phase 2 of the project comprised 247 planning volunteers (22.5%) and 851 on-site volunteers (76.5%). Of the planning volunteer respondents, 108 (44%) were men and 139 (56%) were women. Their average age was 49 years (SD = 12.4; range 18–78 years). The large majority (n = 214, 87%) had previous volunteer experience, while only 33 (13%) did not. Of the on-site volunteer respondents, 316 (37%) were men and 535 (63%) were women. Their average age was 49 years (SD = 13.0; range 18–87 years). As with the planning volunteers, the large majority (n = 684, 84%) of on-site volunteers had previous volunteer experience, while the Games was the first volunteer experience for 126 (16%) in this group. The proportion of planning and on-site volunteers, and men and women volunteers, in the sample corresponded with the population of Canada Games volunteers (23% planning volunteers, 77% on-site volunteers; 39% men, 61% women, with women over-represented in the role of on-site volunteer).

2.3 Instrument

Two survey instruments were developed for the project. 'A survey of 2001 Canada Games volunteers (Part I)' was a pre-Games version that asked volunteers to rate their motives and concerns related to volunteering before they signed on with the Games. Part II of the survey was a post-Games version that asked volunteers to rate the extent to which they experienced certain benefits and costs as a result of their Games involvement, and their intent to volunteer in the future. The items to measure experienced benefits and costs associated with volunteering in Part II of the survey were consistent with the items to measure motives and concerns about volunteering in Part I. This approach ensured that the survey measured 'functionally relevant benefits' (Clary et al., 1998, p. 1524), or those that are consistent with important motives for volunteering in the first place. This approach was extended to the measurement of experienced costs as an extension of initial concerns about volunteering.

The motives/benefits and concerns/costs items were generated from an extensive review of the event volunteer research. At the time of the study, Farrell et al.'s (1998) Sport Event Volunteer Motivation Scale (SEVMS) was the most comprehensive scale in that regard. However, it was determined that the scale did not capture all of the items deemed relevant to event volunteer motives across the literature. This may be due to the fact that it was developed based on a general model of volunteer motivation. Thus, additional sport event-specific motives identified in other studies (Cuskelly, 2000; Janisse & Weese, 2000; Shinde, 1999) were included in the survey; for example, seeing athletes perform, developing a greater appreciation of sport, enjoying privileges of volunteering, and being helpful in providing particular skills and abilities. Items to measure volunteer concerns and costs were also generated from previous research on event volunteers' concerns, expectations, and dissatisfaction regarding their event volunteer experience (Cuskelly, 2000; Elstad, 1996; Shinde, 1999).

The survey of 2001 Canada Games volunteers was pilot tested with a sample of 10 planning volunteers and 40 on-site volunteers with the Games to determine the clarity of the items and instructions, whether any items were missing, and time for completion. Minor adjustments were made, and the final instrument for Part II of the study comprised 31 items representing benefits and 17 items representing costs. Participants were asked to rate on a scale of 1 (completely disagree) to 7 (completely agree) the extent to which they experienced each of the benefits and each of the costs as a result of being a Games volunteer.

There were three measures of future intentions: (1) participants were asked to rate the likelihood that they would volunteer for another major festival or event in their community, on a scale of 1 (definitely would not volunteer) to 4 (not sure) to 7 (definitely would volunteer); (2) they were asked to indicate the extent to which they would want to be involved in a future event in comparison to their degree of involvement with the Games, using a scale of 1 (a lot less) to 4 (the same amount) to 7 (a lot more), while also being given the opportunity to indicate if they would not volunteer again at all; and (3) participants were asked to indicate whether their level of volunteering in the community in general (e.g., with charities, clubs, community service organizations) would change following the Games, using a scale of 1 (will greatly decrease) to 4 (will not change) to 7 (will greatly increase). Space was provided on the survey for participants to add any comments elaborating on their ratings.

2.4 Procedure

A random sample of participants for the entire project was drawn from the total population of 6500 volunteers involved in the Games, which comprised 1500 planning volunteers and 5000 on-site volunteers. For Phase 2 of the project, participants were sent a letter of information and Part II of the survey. Names and addresses were made available through access to the Host Society database through the Volunteer Services Division. Survey packages were mailed to the volunteers' home address with a self-addressed, stamped return envelope 10–15 days after the completion of the Games. One follow-up postcard was mailed two weeks after that (cf. Dillman, 2000). A total of 395 planning volunteers (26%) were sampled from across all positions (Vice Chairs, Sport Leaders, Venue Leaders, Venue Reps, and other volunteer planning committee members). Four surveys were returned due to incorrect address information, and 247 completed surveys were returned for a response rate of 63%. A total of 1171 on-site volunteers (23%) were sampled from the on-site volunteers at 12 Games venues, including at least one in each of the host communities. Of those, 851 returned completed surveys for a response rate of 73%. The overall response rate was 72%.

2.5 Analysis

Principal components analyses with varimax rotation were performed to determine whether the multiple items representing experienced benefits and the items representing experienced costs were able to be reduced to statistically sound and conceptually meaningful factors. Items that loaded higher than .40 on a single factor were retained (Stevens, 2002). Items that loaded higher than .40 on more than one factor were interpreted when the difference between the factor loadings was more than .10 (Stevens, 2002). Any items that did not meet these criteria were eliminated. Cronbach's alpha coefficients were computed to determine the internal consistency of each identified factor; coefficients above .70 were considered to be acceptable (Nunnally & Bernstein, 1994). Separate MANOVAs were conducted to determine whether experienced benefits and experienced costs varied by volunteer type (planning volunteers, on-site volunteers). Given the unequal sample sizes in the data set, Pillai's Trace F-values were examined as they are robust against violations of homogeneity of variance which may occur when cell sizes differ (Tabachnick & Fidell, 2001). A power level of .80 for each test was considered acceptable (Cohen,1992). Given the small proportion of Games volunteers with no previous experience (15%), analysis based on this variable was limited to a consideration of variation in future volunteering intentions.

The data generated from the participants' open-ended comments were transcribed verbatim and categorized according to whether they pertained to experienced benefits, experienced costs, and/or future volunteering intentions. The data supplement the survey results, and references that illuminate and support these findings are presented in the discussion.

3. RESULTS

3.1 Experienced benefits

Six factors representing experienced benefits of volunteering with the Games emerged from the exploratory factor analysis: community contribution (eight items), skill enrichment (four items), connection with sport (four items), privileges of volunteering (five items), positive life experience (three items), and social enrichment (three items). The factor loadings, variance, and Cronbach's alpha coefficients for each factor are presented in Table 1. A total of 27 of the 31 items were retained through the factor analysis. One item did not load on any of the factors, and three items loaded on more than one factor, and thus were eliminated. A one-way MANOVA revealed significant variation in experienced benefits by volunteer type, $F(6,917) = 19.52, p < .001$ (see Table 2). Further univariate ANOVAs indicated that planning volunteers experienced social enrichment, skill enrichment, and privileges to a significantly greater extent than on-site volunteers ($p < .001$).

Table 1. Results of exploratory factor analysis and internal consistency estimates (Cronbach's alpha coefficient) for perceived benefits of event volunteering.

	Factor					
Item	1	2	3	4	5	6
1. Community contribution						
Contributed to community spirit	.74[a]					
Helped community image	.74					
Put something back into community	.71					
Helped make the Games a success	.70					
Felt part of the community	.68					
Helped create a better society	.66					
Did something worthwhile	.54					
Was involved in the Games	.43					
2. Skill enrichment						
My skills were needed		.73				
Was able to use my knowledge, experience		.72				
Acquired, developed new skills		.65				
Gained some practical experience		.61				
3. Connection with sport						
Saw elite athletes			.79			
Watched some events while working			.78			
Mingled with athletes			.65			
Did something to help the athletes perform			.51			
4. Privileges of volunteering						
Got free souvenirs/memorabilia				.76		
Access to places denied the general public				.56		
Was 'behind the scenes' at the Games				.56		
Being a Games volunteer was prestigious				.47		
Worked at a preferred venue/event				.42		
5. Positive life experience						
Developed greater understanding/appreciation of sports					.58	
Did something different, varied my regular activities					.52	
Broadened my horizons					.46	
6. Social enrichment						
Interacted with other volunteers						.73
Made new friends						.62
Worked with different people						.60
Variance (%)	17.5	12.1	10.0	7.3	7.0	6.7
Cronbach's alpha	.90	.85	.76	.74	.77	.74

[a]Loadings < .40 are suppressed.

Table 2. Descriptive (M, SD) and univariate (*F*) results of experienced benefits by volunteer type.

Experienced benefits	Planning volunteers	On-site volunteers	*F*
Social enrichment	6.28 (.87)	5.99 (.99)	14.17***
Community contribution	5.92 (.94)	5.76 (1.01)	4.07*
Positive life experience	5.31 (1.18)	5.43 (1.18)	1.85
Skill enrichment	5.56 (1.08)	4.85 (1.47)	43.84***
Privileges of volunteering	5.19 (1.22)	4.74 (1.32)	12.94***
Connection with sport	4.49 (1.67)	4.46 (1.62)	.03

*$p < .05$, ***$p < .001$

3.2 Experienced costs

Four factors representing experienced costs of volunteering with the Games emerged from the exploratory factor analysis, however, the internal consistency of one of the factors – lack of ability (three items; $\alpha = .60$) – was not acceptable and the factor and corresponding items were eliminated. A fourth item was eliminated because doing so increased the alpha coefficient of the corresponding factor to an acceptable level. The three acceptable factors included: task overload (five items), personal inconvenience (four items), and task underload (four items). The factor loadings, variance, and Cronbach's alpha coefficients for each factor are presented in Table 3. A one-way MANOVA revealed significant variation in experienced costs by volunteer type, $F(3,1033) = 43.63$, $p < .001$ (see Table 4). Further univariate ANOVAs indicated that planning volunteers experienced personal inconvenience and task overload to a significantly greater extent than on-site volunteers ($p < .001$).

Table 3. Results of exploratory factor analysis and internal consistency estimates (Cronbach's alpha coefficient) for perceived costs of event volunteering.

	Factor			
Item	1	2	3	4
1. Task overload				
Undue pressure to perform my tasks well	.73[a]			
Felt taken advantage of by paid staff	.72			
Felt take advantage of by more senior volunteers	.72			
No guidance, support to help me do my tasks	.67			
Not enough other volunteers to help me do tasks well	.61			
2. Personal inconvenience				
Inconvenient for my family/work responsibilities		.71		
Inconvenient to my summer vacation		.69		
Required undue personal expense (out of my own pocket)		.53		
Given too much to do		.52		
3. Task underload				
Assigned boring tasks			.81	
Given too little to do			.80	
Felt undervalued as a volunteer			.62	
Things were generally disorganized			.50	
4. Lack of ability				
Did not have the knowledge/skills to do my tasks				.85
Did not have confidence to do my tasks				.69
Did not receive adequate training for my assignment				.48
Variance (%)	21.6	13.3	13.1	9.5
Cronbach's alpha	.80	.73	.73	.60

[a]Loadings < .40 are suppressed.

Table 4. Descriptive (M, SD) and univariate (*F*) results of experienced costs by volunteer type.

Experienced costs	Planning volunteers	On-site volunteers	*F*
Task underload	2.50 (1.19)	2.77 (1.45)	6.81**
Personal inconvenience	2.60 (1.49)	1.81 (1.06)	81.89***
Task overload	1.87 (1.07)	1.44 (.76)	46.47***

$^{**}p < .01$. $^{***}p < .001$

3.3 Intent to volunteer in the future

Due to the very low proportion of Games volunteers with no previous experience, and in order to avoid further concerns regarding unequal sample sizes, separate one-way MANOVAs were conducted to examine the variation in future volunteering intentions by volunteer type and by previous volunteer experience. To account for these multiple analyses, the alpha level of significance for main effects was set at $p < .001$. Nevertheless, the results must be interpreted with caution. A one-way MANOVA revealed significant variation in future intentions to volunteer by volunteer type, $F(3,1017) = 23.75, p < .001$. Further univariate ANOVAs indicated that on-site and planning volunteers were equally likely to become involved in a future festival or event (M = 5.42, SD = 1.28 vs. M = 5.18, SD = 1.44, respectively). However, they differed significantly in the extent to which they wanted to be involved in comparison to their Games contribution, $F(1,1019) = 69.75, p < .001$; planning volunteers were interested in being involved to the same extent (M = 4.06, SD = 1.47), while on-site volunteers were interested in being involved to a slightly greater extent (M = 4.89, SD = 1.26). On-site volunteers were also more likely to increase their level of volunteering in the community than planning volunteers (M = 4.61, SD = .90 vs. M = 4.37, SD = .83, respectively), $F(1,1019) = 13.78, p < .001$.

Given the significant variation in future intentions between planning and on-site volunteers, separate MANOVAs were conducted to explore whether there was any variation in future intentions among planning volunteers with and without previous volunteering experience, and among on-site volunteers with and without previous volunteering experience. Notably, less than 2% ($n = 20$) of all volunteers indicated they would not volunteer again. Results revealed no significant variation among planning volunteers regarding future intentions. In contrast, significant variation in future volunteering intentions among on-site volunteers with and without previous volunteering experience was indicated, $F(3,740) = 12.37, p < .001$. Further univariate analyses revealed that on-site volunteers with no previous experience were significantly more likely to increase their level of community volunteering than those with previous experience (M = 5.01, SD = .97 vs. M = 4.54, SD = .87), $F(1,742) = 26.72, p < .001$.

Given the significant variation between planning and on-site volunteers in experienced benefits, experienced costs, and future volunteering intentions, separate standard regression analyses were conducted for the two groups. The results for planning volunteers (see Table 5) revealed that perceived benefits and costs explained 33% of the variance in their intent to volunteer for a future event. Specifically, task overload, personal inconvenience, and community contribution were significant predictors. Personal inconvenience was the only significant predictor of the extent of planning volunteers' future event involvement in comparison to their Games contribution (16% variance). Finally, experienced benefits and costs explained 17% of the variance in planning volunteers' expected change in level of community volunteering. Positive life experience and task overload were the significant predictors.

Table 6 presents the results for on-site volunteers and indicates that perceived benefits and costs explained 27% of the variance in their intent to volunteer for a future event. Social enrichment, a positive life experience, community contribution, skill enrichment, and personal inconvenience contributed significantly to that variance. Only 7% of the variance in the extent of on-site volunteers' future event involvement in comparison to their Games contribution was attributed to the experienced benefits and costs measured here. Specifically, task underload, personal inconvenience, and a positive life experience contributed significantly to that variance. Experienced benefits and costs explained 13% of the variance in on-site volunteers' expected change in level of community volunteering. A positive life experience, community contribution, personal inconvenience, and a connection with sport were the significant predictors.

Table 5. Results of standard regression analyses of planning volunteers' experienced benefits and costs, and future volunteer intentions.

Variable	*B*	*SE B*	*β*	*F*	R^2
Volunteer for another event				10.50***	.33
Task overload	–.36	.12	–.26**		
Community contribution	.35	.18	.22*		
Personal inconvenience	–.17	.07	–.17*		
Skill enrichment	.19	.14	.14		
Positive life experience	.15	.11	.12		
Privileges	.11	.10	.09		
Social enrichment	.11	.14	.07		
Connection with sport	.01	.07	.01		
Task underload	.00	.09	.00		
Volunteer to a greater or lesser extent				4.18***	.16
Personal inconvenience	–.33	.08	–.33***		
Community contribution	.21	.20	.13		
Task overload	–.17	.13	–.12		
Task underload	–.14	.10	.11		
Privileges	.12	.11	.10		
Skill enrichment	.12	.15	.09		
Positive life experience	.05	.13	.04		
Connection with sport	.01	07	.01		
Social enrichment	.00	.15	.00		
Change in community volunteering				4.49***	.17
Positive life experience	.22	.07	.30**		
Task overload	–.15	.07	–.18*		
Personal inconvenience	–.06	.05	–.11		
Privileges	.06	.06	.08		
Community contribution	.05	.12	.06		
Task underload	–.05	.06	–.06		
Connection with sport	.03	.04	.05		
Skill enrichment	.02	.09	.02		
Social enrichment	.00	.09	.00		

*$p < .05$, **$p < .01$, *** $p < .001$

Table 6. Results of standard regression analyses of on-site volunteers' experienced benefits and costs, and future volunteer intentions.

Variable	*B*	*SE B*	*β*	*F*	R^2
Volunteer for another event				25.75***	.27
Social enrichment	.22	.06	.16**		
Community contribution	.19	.07	.14*		
Positive life experience	.16	.06	.14*		
Skill enrichment	.12	.05	.12*		
Personal inconvenience	–.11	.05	–.08*		
Task overload	–.08	.07	–.04		
Connection with sport	.05	.03	.05		
Privileges	.02	.05	.02		
Task underload	–.02	.04	–.02		
Volunteer to a greater or lesser extent				5.62***	.07
Personal inconvenience	–.23	.05	–.20***		
Task underload	–.20	.04	–.23***		
Positive life experience	.13	.06	.12*		
Social enrichment	.08	.06	.06		
Skill enrichment	.04	.05	.05		
Connection with sport	.03	.03	.03		
Task overload	–.02	.07	–.01		
Community contribution	.01	.07	.01		
Privileges	.00	.05	.00		
Change in community volunteering				10.76***	.13
Positive life experience	.12	.04	.15**		
Personal inconvenience	–.09	.03	–.10**		
Community contribution	.11	.05	.13*		
Connection with sport	–.05	.02	.09*		
Privileges	.06	.03	.09*		
Task underload	–.03	.03	–.05		
Task overload	–.03	.05	–.03		
Social enrichment	.02	.04	.03		
Skill enrichment	.00	.03	.00		

$^*p < .05$, $^{**}p < .01$, $^{***}p < .001$

4. DISCUSSION

It has been argued that some of the most valuable outcomes of hosting major sport events may be the social benefits that accrue to a community (Ritchie, 2001). These benefits include the legacy of a 'strengthened social structure related to strengthening community volunteerism' (Ritchie, 2001, p. 156). However, there has been little empirical research focused on the volunteer legacy of major sport events. The findings of the current study shed light on the 2001 Canada Games experiences of planning and on-site volunteers, and the influence of those experiences on their intentions to volunteer in the future. Social exchange theory provided a useful framework for considering both positive and negative experiences involved in the volunteering relationship. The findings provide a sense of the volunteer legacy of these Games, and identify aspects of the Games which contributed to that legacy.

In terms of experiences, planning volunteers reportedly worked with different people, made new friends, used and developed new skills, and enjoyed the 'perks' of volunteering to a greater extent than on-site volunteers. This was likely a function of the greater variety in their tasks and their longer involvement with the Games. Notably, both types of volunteers commented that the degree of social enrichment, which was the most experienced benefit, was somewhat of a surprise. They described meeting different people from their own community and across Canada, and working with an incredibly committed group of volunteers as one of the most rewarding aspects of volunteering. All of the participants felt that they experienced the benefits of volunteering from a moderate to fairly high degree (average factor ratings of 4.46 to 6.28 on a seven-point scale). In contrast, the costs of volunteering were experienced to a limited degree, if at all (average factor ratings of 1.44 to 2.77 on a seven-point scale). Nevertheless, some significant variations were observed such that, in comparison to on-site volunteers, planning volunteers reportedly experienced greater inconvenience to their work, family, and/or summer vacation, and felt undue pressure to perform well, taken advantage of by paid staff and/or more senior volunteers, and a lack of sufficient volunteers to do the work. Again, these costs were fairly low. Related research has also revealed that benefits are experienced by event volunteers to a greater extent than costs, and particularly social enrichment through networking and a positive group environment (e.g., Elstad, 1996; Farrell et al., 1998; Larocque et al., 2002). The current study builds on this knowledge by further distinguishing the respective experiences of planning and on-site volunteers.

Of particular interest in this study was the impact of these experiences on planning and on-site volunteers' future intentions. It is notable that essentially all of the participants (98.2%) indicated they would volunteer for another major event, and both planning and on-site volunteers were fairly likely to do so. This is a positive testament to the volunteer legacy of the Games, and is consistent with previous research that has demonstrated a strong willingness of volunteers to get involved in another event (Downward & Ralston, 2006; MacLean & Hamm, 2007). Another positive legacy is that both planning and on-site volunteers anticipated increasing at least slightly their level of community volunteering, with on-site volunteers expecting to increase their involvement significantly more than planning volunteers. Further, the findings suggested that on-site volunteers with no previous experience were more likely to increase their community volunteering than those with previous experience. The Games seemingly opened them up and turned them on to volunteering. These findings contrast those of Downward and Ralston (2006) who found no variation in future intentions among those with previous volunteer experience and those with no experience. Notably, the distinction here is significant among on-site volunteers whom Downward and Ralston did not distinguish in their analysis.

Results of the regression analyses for planning volunteers revealed that, while the opportunity provided by the Games to make a difference in the community was a significant determinant of their intent to volunteer for another major event, having too much to do without sufficient support and being personally inconvenienced by the experience detracted from the likelihood that they would get involved again. The effect of each of these factors was small (Cohen, 1992), yet, altogether the experienced benefits and costs explained 33% of the variance in planning volunteers' intent to get involved in a future event. Other factors can certainly be expected to influence this decision; however, the findings shed light on the impact of particular experiences. Notably, costs were a more prevalent factor than benefits in predicting planning volunteers' future intentions. Social exchange theory purports that individuals will engage in activities where they expect to receive benefits while incurring minimal costs (Homans, 1958; Thibault & Kelley, 1959). The findings here suggest that the benefits of event volunteering may be seen as a given to planning volunteers while particular costs may be more variable and thus weigh particularly heavily in terms of whether they would become involved in a future event; highlighting the importance of these costs to the violation of the reciprocal exchange relationship (cf. Zafirovski, 2005). Indeed, the cost of personal inconvenience was the sole significant predictor of whether planning volunteers would become involved to a greater or lesser extent in a future event, having a moderate effect on this variable (cf. Cohen, 1992). Thus, to the extent that planning volunteers considered their Games contribution to have been an inconvenience to their job, their family, or their leisure/vacation time, they would be unlikely to maintain, much less increase, their involvement in future events. This is not surprising, since their given level of involvement would have been a factor in the perceived inconvenience of the experience.

In contrast, to the extent that planning volunteers had a positive life experience in terms of doing something different and expanding their horizons they expected to increase their level of community volunteering in general. This is consistent with the study by Downward and Ralston (2006) who found that the personal development experience significantly predicted event volunteers' increased interest in general volunteering. In the current study, the Games experience may have opened their eyes to other possibilities for personal growth, which they could pursue through community volunteering. Being part of a unique family event that exuded incredible spirit was a common example given of the positive life experience of being involved. Nevertheless, these intentions were significantly tempered by the extent to which planning volunteers experienced task overload during the Games, although the effect was relatively milder. Thus, undue pressure to perform well without sufficient guidance or support was a factor in planning volunteers' intentions to extend themselves again in another major event, or in the community in general. The more comprehensive consideration of positive and negative experiences in the current study expands our understanding of factors that impact the likelihood of future volunteering.

Results of the regression analyses for on-site volunteers revealed that positive interactions with fellow volunteers, Games participants, and spectators, making a contribution to one's community, and developing one's skills while doing something different increased the likelihood that they would volunteer for another event in the future. Downward and Ralston (2006) also found personal development, including developing new skills and having a new challenge, to be a factor in event volunteers' willingness to be involved in another major event. However, to the extent that on-site volunteers in the current study experienced conflict with family and/or work responsibilities, they would be less likely to be involved. Although the effect of each factor was small, it is notable that experienced benefits seemed to be a greater force than costs in terms of predicting event volunteering in the future for this group. This is in contrast to the seemingly greater influence of experienced costs on the future intentions of planning volunteers. The findings highlight variations in the importance of various socioemotional outcomes (cf. Cropanzano & Mitchell, 2005) in the exchange relationship for different types of volunteers. Nonetheless, it is not surprising that for both types of volunteers experienced personal inconvenience was a condition of future involvement, as it may be expected that one would only volunteer if family and work responsibilities permit it. Personal inconvenience was also a significant predictor of the extent to which on-site volunteers would become involved in a future event, along with task underload. Thus, being personally inconvenienced, as well as not having enough to do or being given boring tasks, detracted from their willingness to get more involved in a future event. Again, volunteers may not be expected to do more in the future if they previously experienced conflict with family or work commitments. However, the findings also indicate that being given too little to do after one has made a commitment to be involved may detract from their willingness to invest even more time in a future event. Indeed, on-site volunteers noted their disappointment and frustration with not being given enough work to do, particularly because too many volunteers were assigned for a given amount of work.

Although several benefits of volunteering for the Games significantly contributed to the likelihood that on-site volunteers would increase their level of community volunteering, the effects were small. As well, being personally inconvenienced by the Games experience only mildly detracted from that. Thus, although there appears to be a legacy of further community volunteering in general among on-site volunteers, the Games experience had little effect on that. The weak connection may be a function of the short-term and task-specific nature of their Games involvement, in contrast to broadly defined community volunteering in general.

It is notable that, although social enrichment was experienced to a high degree and reported to be particularly rewarding, it was only a mild predictor of future event volunteering, and only for on-site volunteers. This finding seems to contrast Brennan's (2005) suggestion that interactions among volunteers and contributing to a common cause will engender further voluntary action. Instead, volunteers tended to consider their contribution to the community through involvement in the event, as well as having a positive life experience, as particularly meaningful to their future volunteering. It is possible that the social environment provided the mechanism for volunteers to feel like they made a contribution and had a different life experience. The meaning of social enrichment to the volunteer experience, and future voluntary action, should be examined further.

Overall, the findings suggest that the experience of volunteers with a given event can have a significant impact on future volunteer behavior in other events and in the community in general. A legacy of volunteerism can be a meaningful element of event hosting. Therefore, it is essential that event organizers give strong consideration to the most important aspects of event volunteering for planning and on-site volunteers, and ensure that a positive experience with minimal costs will pay off for the event, the volunteers, and the community, both now and in the future.

4.1 Implications for major sport event policy and management

The findings have several implications for strategies to ensure a positive experience for both planning and on-site volunteers. Although experienced benefits were generally less influential than costs with regard to planning volunteers' future intentions, it is still incumbent upon event organizers to ensure these individuals are able to at least feel they have made a contribution to their community and have had a positive life experience through doing something different. The former may be optimized by ensuring that planning volunteers, many of whom work 'behind the scenes,' can see or hear the impact they are having on the event and their community; this may occur through observing events and direct feedback from event participants and/or spectators. Major sport events may provide an opportunity for planning volunteers to try something new (particularly if they have not been involved in a high profile or sport event in the past), and thus look further for different opportunities in the community. Thus, the opportunity to do something different and expand one's horizons should be recognized as a valued part of the planning volunteers' involvement.

Nonetheless, the volunteer legacy associated with planning volunteers appears to be relatively more dependent on their experiences of task overload and personal inconvenience. Thus, organizers should recognize the potential impact of conflicts between event volunteering and work and family commitments on future involvement, and work with planning volunteers to try to minimize these conflicts before and during an event. Effective communication, scheduling meetings well in advance, and ensuring enough volunteers to share the load are potential considerations. These may also help to minimize a sense of task overload, which was a significant predictor of future volunteering. Planning volunteers described this as too high expectations for task completion with insufficient resources to ensure success. This experience may be inherent in major event organizing, particularly when the event is being held in a community for the first time. Organizers may not have a solid grasp of what is required and, as a result, there are continually increased expectations of the volunteer workforce with no greater resources (both financial and human) to support that. This likelihood should be acknowledged by all those involved in planning a major sport event, so that flexibility may be built in to allow for increased expectations and the provision of sufficient support when that occurs. Otherwise, planning volunteers may feel that organizers are merely taking advantage of their commitment to helping with the event. Again, this perception can have important ramifications for their likelihood of volunteering in the future.

In contrast, on-site volunteers' future volunteering intentions were more a function of the benefits they experienced during the Games. Event organizers should ensure that this group of volunteers has an opportunity for social enrichment through working closely with other volunteers, as well as meeting event participants and visitors if possible. Where it is not possible to ensure that on-site volunteers have contact with visitors, given the nature of their assignment, it may be even more important to promote a positive working environment of comradeship and cooperation among the volunteers. In addition, event organizers should recognize the connection between on-site volunteers feeling they contributed to the community and had a new life experience and the likelihood that they will be involved in the future. Again, it is incumbent upon organizers to ensure that all on-site volunteers, regardless of their task assignment, have an opportunity to feel part of the community through helping make the Games a success while broadening their horizons through being involved. The influence of task underload for on-site volunteers' future intentions, and their sense that having too little to do was a function of too many volunteers scheduled, suggests it is important to ensure volunteers have enough to do during the time they are scheduled to work. This requires careful consideration of the work to be done, and effective scheduling of volunteers to do that work. There may be a tendency to focus on the here and now of staging a successful event; however, the findings of this study show that what happens here and now, in the eyes of the volunteers, will have ramifications for the volunteer legacy of the event. The volunteer experience itself must be an additional layer for consideration in event management.

4.2 Limitations to the study and recommendations for future research

There are several limitations to the study that should be acknowledged, as well as recommendations for future research. First, the results of this study pertain to the volunteers of the 2001 Canada Games, and thus generalizing from the findings is limited to that population or similar populations of planning and on-site major sport event volunteers. Second, it is possible that the results reflect a response bias to the extent that individuals who were more positive or enthusiastic about their Games experience may have been more likely to respond. Third, there are inherent limitations in asking about future behavioral intentions without reference to a specific event or activity (Sheeran, 2002). In this study, participants were asked about their intent to volunteer for any other event, and to volunteer in the community in general. This may have inherently reduced the strength of the relationships between the volunteers' experiences and their future intentions. Future research may attempt to measure more specific intentions for volunteering. Fourth, although the measure of behavioral intention is appropriate for understanding future behavior (Sheeran, 2002; Tett & Meyer, 1993), research would ideally measure volunteer legacy in terms of the actual percentage of major sport event volunteers who went on to volunteer for another event in their community, the extent of their involvement in comparison to previous events, and whether they had changed their level of volunteering in the community. Downward and Ralston (2006) included a few measures of actual changes in volunteering subsequent to the 2002 Commonwealth Games, and found very little direct effect of event volunteers' experiences on subsequent volunteering; however, they also noted the need to increase the explanatory power of their research. The time lag in asking about one's event experience almost one year later may have reduced the strength of the relationship to actual volunteering at that point. Longitudinal research that tracks volunteers' perceptions, intentions, and actual behavior at different points post-event is recommended to better understand the link between event volunteer experiences and future volunteering. It would be of particular interest to determine whether first time volunteers continued their involvement, as this may be the true mark of the volunteer legacy of any event.

A fifth recommendation pertains to the very small proportion of volunteers with no previous experience in the current study, which precluded anything more than a basic analysis of their future intentions. Future research may focus on attaining as large a sample as possible of event volunteers with no previous experience so that their experiences and future intentions may be better understood, and compared to their veteran counterparts. Finally, to further enhance our understanding of the volunteer legacy of major sport events, future research should identify factors that may explain variation in those event volunteers' experiences that predict future volunteering; for example, there may be variation in experienced benefits and costs according to the type of volunteer task assignment, venue assignment, and hours worked. Also, following from Downward and Ralston (2006), benefits and costs experienced by volunteers, and future volunteering intentions, may vary by personal demographics, such as age, gender, employment status, and family status (i.e., dependents). Future investigations may provide further insight into factors that shape voluntary action in the community, and the role of major sport events in promoting that legacy.

REFERENCES

1 2001 Alliance London Jeux du Canada Games. 2001. General orientation manual, London, ON: 2001 Alliance London Jeux du Canada Games.

2 Blau, P. 1994. Structural contexts of opportunities, Chicago: University of Chicago Press.

3 Brennan, M.A. 2005. Volunteerism and community development: A comparison of factors shaping volunteer behavior in Irish and American communities. Journal of Volunteer Administration, 23(2): 20–27.

4 Clary, E.G., Snyder, M., Ridge, R.D., Copeland, J., Stukas, A.A.Haugen, J. 1998. Understanding and assessing the motivations of volunteers: A functional approach. Journal of Personality and Social Psychology, 74: 1516–1530. ,

5 Cnaan, R.A. and Goldberg-Glen, R.S. 1991. Measuring motivation to volunteer in human services. Journal of Applied Behavioral Sciences, 27: 269–284.

6 Cohen, J. 1992. A power primer. Psychological Bulletin, 112: 155–159. ,

7 Costa, C.A., Chalip, L., Green, B.C. and Simes, C. 2006. Reconsidering the role of training in event volunteers' satisfaction.Sport Management Review, 9: 165–182.

8 Crompton, J. 2004. Beyond economic impact: An alternative rationale for the public subsidy of major league sports facilities. Journal of Sport Management, 18: 40–58.

9 Cropanzano, R. and Mitchell, M.S. 2005. Social exchange theory: An interdisciplinary review. Journal of Management, 31: 874–900. ,

10 Cuskelly, G. Predicting the behavioral dependability of volunteers at major sport events: An application of the theory of planned behavior. Paper presented at the North American Society for Sport Management Conference. Colorado Springs, CO. May.

11 Cuskelly, G., Hoye, R. and Auld, C. 2006. Working with volunteers in sport, London: Routledge.

12 Dillman, D.A. 2000. Mail and internet surveys: The tailored design method , (2nd ed.), New York: John Wiley.

13 Downward, P. and Ralston, R. 2006. The sports development potential of sports event volunteering: Insights from the XVII Manchester Commonwealth Games. European Sport Management Quarterly, 6: 333–351.

14 Elstad, B. 1996. Volunteer perceptions of learning and satisfaction in a mega-event: The case study of the XVII Olympic Winter Games in Lillehammer. Festival Management and Event Tourism, 4: 75–83.

15 Farrell, J.M., Johnston, M.E. and Twynam, G.D. 1998. Volunteer motivation, satisfaction, and management at an elite sporting competition. Journal of Sport Management, 12: 288–300.

16 Gladden, J.M., McDonald, M.A. and Barr, C.A. 2005. "Event management". In Principles and practice of sport management , (2nd ed.), Edited by: Masteralexis, L.P., Barr, C.A. and Hums, M.A. 272–294. Sudbury, MA: Jones and Bartlett.

17 Gratton, C., Shibli, S. and Coleman, R. 2005. Sport and economic regeneration in cities. Urban Studies, 42: 985–999.,

18 Gratton, C. and Taylor, P. 2000. Economics of sports and recreation, London: Taylor & Francis.

19 Guillet, E., Sarrazin, P., Carpenter, P.J., Trouilloud, D. and Cury, F. 2002. Predicting persistence or withdrawal in female handballers with social exchange theory. International Journal of Psychology, 37: 92–104. ,

20 Hall, M., Lasby, D., Gumulka, G. and Tryon, C. 2006. Highlights from the 2004 Canada survey of giving, volunteering, and participating, Ottawa, ON: Statistics Canada.

21 Hiller, H.H. 2006. Post-event outcomes and the post-modern turn: The Olympics and urban transformations. European Sport Management Quarterly, 6: 317–332.

22 Homans, G. 1958. Social behavior as exchange. American Journal of Sociology, 63: 597–606. ,

23 Howard, D.R. and Crompton, J.L. 2004. Tactics used by sports organizations in the United States to increase ticket sales.Managing Leisure, 9: 87–95.

24 Janisse, M. and Weese, W.J. 2000. Volunteer profiles, motivation and satisfaction: A study of the 1998 Detroit Grand Prix Association, Canada: Unpublished manuscript, The University of Windsor.

25 Larocque, L., Gravelle, F. and Karlis, G. Volunteerism and serious leisure: The case of the Francophone Games. Paper presented at the proceedings of the 10th Canadian Congress on Leisure Research University of Alberta. Edmonton, Canada. May. pp.181–183.

26 Lynch, B. 2001. Volunteers 2000: The Sydney experience, Sydney: SOCOG.

27 MacLean, J. and Hamm, S. 2007. Motivation, commitment, and intentions of volunteers at a large Canadian sporting event.Leisure/Loisir, 31: 523–556.

28 Matheson, V.A. and Baade, R.A. 2006. Padding required: Assessing the economic impact of the Super Bowl. European Sport Management Quarterly, 6: 353–374.

29 Misener, L. and Mason, D.S. 2006. Creating community networks: Can sporting events offer meaningful sources of social capital?. Managing Leisure, 11: 39–56.

30 Misener, L. and Mason, D.S. 2008. Urban regimes and the sporting events agenda: A cross-national comparison of civic development strategies. Journal of Sport Management, 22: 603–627.

31 Nunnally, J.C. and Bernstein, I.H. 1994. Psychometric theory , (3rd ed.), New York: McGraw-Hill.

32 Ritchie, J.R.B. 2001. Turning 16 days into 16 years through Olympic legacies. Event Management, 6: 155–165.

33 Sheeran, P. 2002. "Intention-behavior relations: A conceptual and empirical review". In European review of social psychology, Edited by: Hewstone, M. and Stroebe, W. Vol. (12), 1–36. Chichester: Wiley.

34 Shinde, M. Volunteer satisfaction for Nagano Winter Olympics. Paper presented at the North American Society for Sport Management Conference. Vancouver, BC. May.

35 Smith, D.H. 1981. Altruism, volunteers, and volunteerism. Journal of Voluntary Action Research, 10(1): 21–36.

36 Solberg, H.A. 2003. Major sporting events: Assessing the value of volunteers' work. Managing Leisure, 8: 17–27.

37 Sterken, E. 2006. Growth impact of major sporting events. European Sport Management Quarterly, 6: 375–390.

38 Stevens, J. 2002. Applied multivariate statistics for the social sciences , (4th ed.), Mahwah, NJ: Lawrence Erlbaum.

39 Tabachnick, B.G. and Fidell, L.S. 2001. Using multivariate statistics , (4th ed.), Needham Heights, MA: Allyn & Bacon.

40 Tett, R.P. and Meyer, J.P. 1993. Job satisfaction, organizational commitment, turnover intention, and turnover: Path analyses based on meta-analytic findings. Personal Psychology, 46: 259–293.,

41 Thibault, J.W. and Kelley, H.H. 1959. The social psychology of groups, New York: Wiley.

42 Twynam, G.D., Farrell, J.M. and Johnston, M.E. 2002/2003. Leisure and volunteer motivation at a special sporting event.Leisure/Loisir, 27: 363–377.

43 Volunteer, BC. 2008. Volunteering for the 2010 Olympic Winter Games Retrieved January 7, 2009, fromhttp:// www. volunteerbc.bc.ca/olympics/index.html

44 Waitt, G. 2003. Social impacts of the Sydney Olympics. Annals of Tourism Research, 30: 194–215.,

45 Weiss, M.R. and Stevens, C. 1993. Motivation and attrition of female coaches: An application of social exchange theory.Sport Psychologist, 7: 244–261.

46 Yan, Z. and Chen, L. 2008. Smiling volunteers win another Olympic gold for Beijing. Chinaview, Retrieved January 7, 2009, from http://news.xinhuanet.com/english/2008-08/14/content_9321992.htm

47 Zafirovski, M. 2005. Social exchange theory under scrutiny: A positive critique of its economic-behaviorist formulations.Electronic Journal of Sociology, Retrieved March 11, 2009, from http://www.sociology.org/ content/2005/ tier2/ SETheory.pdf

CHAPTER 8

Models of Volunteer Management: Professional Volunteer Program Management in Social Work

Jeffrey L. Brudney[a*] **& Lucas C.P.M. Meijs**[b]

[a] Department of Public and International Affairs, University of North Carolina Wilmington, Wilmington, North Carolina, USA

[b] Rotterdam School of Management, Erasmus University, Rotterdam, The Netherlands

ABSTRACT

Several trends are leading to increased and broader involvement of volunteers in social work practice. As a consequence, social workers need to be able to manage volunteers in different settings, based on organizational/program factors and characteristics of the volunteers. Contemporary research on volunteer management can be divided into universalistic and contingency approaches. This article presents an overview of leading concepts in both perspectives and offers recommendations for social workers to select appropriate approaches to manage volunteers professionally across different contexts.

KEYWORDS

contingency, management, practice, universal, volunteer

1. INTRODUCTION

Across diverse settings, social work practice will involve working with volunteers and oftentimes managing their activities (Sherr, 2008). The involvement of social work professionals with volunteers ranges from serving as the administrator of an agency's volunteer program, to cooperating with independent or self-help/mutual aid volunteer groups, to mobilizing family members, friends, and neighbors of clients to assist in intervention or treatment, to helping individuals re-engage in social life through volunteering. Social workers are involved with a variety of volunteers, from younger to older, from highly skilled to unskilled, and across ethnic/racial lines. In sum, social work professionals encounter volunteers in various forms and in countless settings. This article offers a broad picture of volunteer management approaches and tools with the aim of increasing social workers' ability to engage and work with volunteers most effectively.

The mainly practitioner literature on volunteer administration has recommended best practices to aid social workers and others for this purpose, such as preparing job descriptions for volunteers, matching volunteers' interests and capabilities to unpaid organizational positions, training and orienting volunteers, and having policies and procedures (for example, McCurley & Lynch, 2011; Ellis, 2010). However, academic studies attempting to confirm the existence and use, let alone the effectiveness, of best practices in volunteer administration have been scarce, and the results inconclusive (Hager & Brudney,2011; Cuskelly, Taylor, Hoye, & Darcy, 2006). Although problems in the conceptualization and measurement of volunteer performance may be responsible at least in part for these disparate findings (Safrit, 2010), we suspect that the issues involved in confirming the impact of best practices have roots in more fundamental differences distinguishing volunteer programs. From this perspective, the lack of empirical confirmation of general volunteer management best practices may stem from the great diversity of volunteer programs and management practices that exist to meet a variety of societal needs. If one management approach is not sufficient to deal with the diversity of volunteering, measuring using this one management approach will not lead to confirmation of its use.

In our view, the field of volunteer management could benefit from the use of contingency approaches, similar to what has been utilized to address paid staff management in strategic human resource management (Delery & Doty, 1996). From this point of view, universal "best practices" are valid only within a specific contingency or set of conditions. In this article we present universal and contingency approaches to volunteer management, and explain under what circumstances the different approaches are likely to be most useful.

1.1 The Issue and Current Literature

The need for collaboration between volunteers and social workers has long been a topic of debate within the social work profession.

Writing in the early 1960s, Becker (1964) cited personnel shortages as the primary reason for the continued and expanded use of volunteers. In the early days of social work, volunteers, known as "friendly visitors" (p. 58), laid the foundation upon which current social work practice has been built. Typically white, middle- and upper-class women, friendly visitors embraced volunteering as both a civic duty and an opportunity to fill their leisure time with more meaningful activities. Over time, this practice gave way to the settlement house movement and, later, a growing disapproval of friendly visitors, who were seen as condescending and unprepared to handle the challenges of working with the poor. As social work developed into a field separate from the work of charitable organizations, paid caseworkers began to supersede volunteers in working directly with clients, while the role of the volunteers diminished to more menial tasks with minimal responsibility (p. 59–69). In spite of, or perhaps because of, their longtime partnership with volunteers, social workers sought to distance themselves from the seemingly patronizing practices of the past by greatly limiting volunteer involvement with casework.

Sherr (2008) discusses the historical relationship between social work and volunteers, and further examines the emergence and entrenchment of the tension between them. Like Becker (1964), Sherr identifies the quest for professional status as the primary impetus for restricting the responsibilities of volunteers in social work practice. Although in the past social work pioneers such as Mary Richmond envisioned professional social workers as managers whose primary concern was to coordinate and sustain a diverse and productive group of volunteers, Abraham Flexner's startling assessment of social work as "not a profession" at the National Conference of Charities and Corrections in 1915 led many to believe that the quickest way to achieve professional status was to reduce the role of volunteers and establish firm boundaries between volunteerism and social work. Though rendered a century ago, this preoccupation may still limit social work's ability to carry out its mission of serving those in need and enacting social change (p. 69–70).

Looking ahead to the future of social work necessitates looking back to the foundations of the field. Although some aspects of past practice cannot and should not be applied to the present day, Sherr (2008) suggests that contemporary social workers can learn many lessons from their predecessors of the 19th and early 20th centuries about organizing and employing volunteers. Examining the work of Jane Addams from a new perspective, he argues that her success as a leader in social work was derived primarily from her ability to inspire people from different backgrounds to identify common interests and to combine their energy into a catalyst for social change. Addams was not only a masterful organizer but also an effective manager who recognized the potential in others and maximized the human capital that she found through numerous social clubs and meetings to achieve higher organizational goals. In cultivating a broad and diverse volunteer base, Addams increased the longevity and impact of her work and passed along her vision for social improvement to many others (Sherr,2008, p. 49–61).

Jane Addams exhibited a great affinity for networking. This approach is a part of the historical legacy of social work, but it must also play a role in its future development, especially in the education and training of prospective social workers. Sherr (2008) writes,

Just as social workers learn to be conscious and deliberate in how they develop and maintain trusting relationships in clinical practice, they must do the same in strategically developing a diverse network of genuine relationships with people who may become volunteer partners. (p. 115)

In seeking to build their own diverse networks of volunteers, individual social workers should look to voluntary organizations as potential sources of volunteers and resources in the ongoing quest for social change. According to Theilen and Poole (1986), "forming, holding membership in, or collaborating with voluntary associations" (qtd. in Sherr, 2008, p. 116) can expose social workers firsthand to this vast and currently underutilized resource.

Another potential source of volunteers that social workers should consider is former clients. In implementing "long-range intervention strategies" (Sherr, 2008, p. 126), social workers often encounter difficulties in ensuring program effectiveness due to the lack of adequate staffing. Volunteers, especially those who have first-hand experience with social work strategies through their own participation in similar programs, offer an affordable and readily available solution to this problem. Likewise, former clients help to correct for the shortcomings of the volunteer efforts of the past by bringing even greater diversity and insight to the table.

Vinton (2012) describes several factors that explain why contemporary social work organizations will need to focus more on the management of volunteers. Especially as a consequence of the pressure on budgets under prevailing economic conditions, the general retreat of the governmentally financed welfare state, the growing demand for social services, and changes in the availability of volunteers, social workers across all policy fields and organizational contexts and cultures will feel pressured to involve more volunteers. Although working with and managing volunteers is not new to social work professionals (see, for example, Schwartz, 1984; Perlmutter & Cnaan, 1994; Netting, Nelson, Borders, & Huber, 2004), the use of volunteers, and their effective management, is becoming more critical. The need for professional practice is acute as volunteers and the modes of their involvement grow more complex (Wilson, 2012), and the conceptualization of volunteering embraces this fluidity (Brudney & Meijs, 2009). As Vinton (2012) puts it, "A well-managed volunteer program can mean that services to clients do not necessarily have to be cut" (p. 146). Overcoming the tension between social work and volunteering that Sherr (2008) describes has become even more critical under the current circumstances.

Historically, social work grew from the efforts of a mostly homogenous group of well-meaning and hardworking, if technically untrained, volunteers. Professionally, social work sought to distance itself from this past by greatly limiting the role of volunteers to one of support, rather than direct involvement in casework. However, this tension cannot and should not be sustained. As the profession comes to another crossroads in its development, demands for increased staffing and support can be met through the integration of a diverse network of volunteers with a wider scope of responsibilities. While professional social workers should retain their current function as facilitators and enablers of client welfare, they should be open to embracing an expanded role as managers of volunteers as well.

The practitioner literature on volunteer management holds prevailing assumptions regarding universal "best practices." As Rochester (1999) explains, this approach embodies "the view that 'volunteering is volunteering is volunteering,' that what is being measured or described is essentially the same activity regardless of context" (p. 10). Increasingly, however, this literature encounters academic research that questions a universalistic approach to such a broad activity and instead adopts a "contingency" approach; here (universal) best practices are applicable or appropriate but only under certain conditions (Hager & Brudney, 2011). Thus, an effective administrative approach to volunteer management should be contingent on a variety of factors, such as the type or role of volunteers to be managed, the characteristics of the volunteer program, and the culture of the organization (for example, Macduff, Netting, & O'Connor, 2009; Rehnborg, 2009; Rochester, 1999). We discuss several of the main contingency approaches in this article, but first we turn to some highly influential universal volunteer management models.

1.2 Universalistic Volunteer Management

Among those with a universalistic perspective on volunteer management is Susan Ellis, one of the leading trainers in volunteer administration. In her 2010 book *From the Top Down*, Ellis asserts,

The skills of volunteer administration are generic and apply to all settings. They are also amazingly universal. We have presented sessions in twenty-six countries on six continents; the context varies from culture to culture, but the principles always apply (p. viii).

Her book provides a "Volunteer Involvement Task Outline" framed by nine necessary functions of successful volunteer programs: planning and administration, volunteer work design, recruitment, interviewing and screening, orientation and training, supervision and liaison support, ongoing motivation and recognition, impact evaluation, and recordkeeping and reporting (Ellis, 2010, pp. 269–273). She also presents a tenth category, "Other Responsibilities (as applicable to each organization)" (p. 273), and even though the heading appears more conditional than universal, the list of prescriptive actions that follows does not take into account any contingency factors.

Scholars and practitioners with a universalistic outlook advance a "one size fits all" conception of volunteer management. However, disagreement exists even among "universalists" regarding the particulars of the universal model—which implies the need for and potential of contingency approaches. The UPS Foundation, in partnership with the Association for Volunteer Administration and the Points of Light Foundation, offers another universalistic model of volunteer management in "A Guide to Investing in Volunteer Management" (UPS Foundation, 2002). Included in the guide is a 23-item checklist of "Elements of Volunteer Resource Management" (p. 15), which are presented as basic requirements for the successful involvement of volunteers. As shown in Table 1, each of the components seems to fall into one of three areas: policy and program structure, management, or program evaluation. The UPS model is grounded on the assumption that the management of volunteers requires a specific set of policies, management skills, and program evaluation techniques across all organizations.

Table 1 Elements of Volunteer Resource Management

1. Written statement of philosophy related to volunteer involvement
2. Orientation for new paid staff about why and how volunteers are involved in the organization's work
3. Designated manager/leader for overseeing management of volunteers agency-wide
4. Periodic needs assessment to determine how volunteers should be involved to address the mission
5. Written position descriptions for volunteer roles
6. Written policies and procedures for volunteer involvement
7. Organizational budget reflecting expenses related to volunteer involvement
8. Periodic risk management assessment related to volunteer roles
9. Liability insurance coverage for volunteers
10. Specific strategies for ongoing volunteer recruitment
11. Standardized screening and matching procedures for determining appropriate placement of volunteers
12. Consistent general orientation for new volunteers
13. Consistent training for new volunteers regarding specific duties and responsibilities
14. Designated supervisors for all volunteer roles
15. Periodic assessment of volunteer performance
16. Periodic assessments of staff support for volunteers
17. Consistent activities for recognizing volunteer contributions
18. Consistent activities for recognizing staff support of volunteers
19. Regular collection of information (numerical and anecdotal) regarding volunteer involvement
20. Information related to volunteer involvement shared with board members and other stakeholders at least twice annually
21. Volunteer resources manager and fund development manager work closely together
22. Volunteer resources manager included in top-level planning
23. Volunteer involvement linked to organizational outcomes

Source: UPS Foundation (2002, p. 15).

Machin and Paine (2008) also present a universalistic framework in their "Management Matters: A National Survey of Volunteer Management Capacity." They assess the extent to which "recognised elements of good practice" (Machin & Paine,2008, p. 5) direct volunteer management, "as outlined in ten indicators for achieving Investing in Volunteers, the UK quality standard for all organisations which involve volunteers in their work" (p. 28). As shown in Table 2, the best practices are based on nine indicators for achieving "quality" and reflect the view that they must be in place to yield success in managing volunteers (Investing in Volunteers, 2010).

Table 2 UK Quality Standard for Organizations that Involve Volunteers

Nine Indicators of Quality
1. There is an expressed commitment to the involvement of volunteers and recognition throughout the organization that volunteering is a two-way process that benefits volunteers and the organization.
2. The organization commits appropriate resources to working with all volunteers, such as money, management, staff time, and materials.
3. The organization is open to involving volunteers who reflect the diversity of the local community and actively seeks to do this in accordance with its stated aims.
4. The organization develops appropriate roles for volunteers in line with its aims and objectives, which are of value to the volunteers.
5. The organization is committed to ensuring that, as far as possible, volunteers are protected from physical, financial, and emotional harm arising from volunteering.
6. The organization is committed to using fair, efficient, and consistent recruitment procedures for all potential volunteers.
7. Clear procedures are put into action for introducing new volunteers to their role, the organization, its work, policies, and practices, and relevant personnel.
8. The organization takes account of varying support and supervision needs of volunteers.
9. The whole organization is aware of the need to give volunteers recognition.

Source: Investing in Volunteers (2010).

In sum, the literature forwards universalistic practices for volunteer management, which do not vary by any organizational or volunteer contingencies identified by the authors. Thus, the management practices would apparently apply to all organizational situations and circumstances of social work practice, whether an all-volunteer organization where the professional social worker may be the only paid staff member, a volunteer department or program in a well-established social work agency, an informal neighborhood group involved in providing food to homeless people, a group of parents volunteering their time in a local school or neighborhood improvement effort, or an advocacy group of concerned citizens working toward a social justice cause. Although the authors may allow for nuance in implementation, the conditions for differential application of the practices are implied, rather than stated. The conditional approaches to volunteer management that would guide social workers are much more specific and claim to be helpful in creating volunteer management programs that work under various circumstances.

1.3 Conditional Volunteer Management

Paull (2002) summarizes the contingency approach that "a 'one size fits all' approach is inappropriate for volunteer management ... 'it really does depend on the type of organization'" (p. 25). He suggests that effective volunteer management must adapt according to such factors as the presence of paid staff in the organization, organizational size, and grassroots membership. Meijs and Ten Hoorn (2008) agree on the following:

There is simply no best way of organizing volunteers, neither in volunteer run organizations, in government organizations, in nonprofit organizations with mostly paid staff, nor in businesses. Volunteering, volunteers and the way they are organized and managed differs from context to context. (p. 29).

Volunteer management literature that adopts a conditional perspective emanates from the proposition that circumstances or conditions must dictate management practices in social work and other fields. These models can become quite complex, as shown by Studer and Von Schnurbein (2013). Although researchers have focused on a variety of conditions that may affect volunteer management the most frequent contingency factors in the conditional volunteer management literature are either volunteer-focused or program/organization-focused.

1.4 Volunteer-Focused

Based on a review of the relevant literature, Rochester (1999) fixes on the role of the volunteer in his "One Size Does Not Fit All: Four Models of Involving Volunteers in Small Voluntary Organizations." As displayed in Table 3, he proposes four models of volunteer involvement service delivery, support role, member/activist, and co-worker (p. 12).

Table 3 Four Models of Volunteer Involvement

	Type of Model			
	Service Delivery Model	*Support Role Model*	*Member/Activist Model*	*Co-worker Model*
Relationship to Volunteer				
Role of volunteer	Most of work done by volunteer	Volunteer supplement work of paid staff	All positions held by volunteers	Unclear distinctions between volunteer and paid staff
Recruitment of volunteer	Specific recruitment based on volunteer ability	Volunteer recruited to take a non-operational role	Volunteer's purpose in organization is self-defined	Volunteer's purpose in organization is self-defined
Volunteer motivation	Potentially relevant to paid employment	Feeling of doing good	Volunteer involved for personal growth and development	Develop or maintain a particular service
Volunteer management	"Workplace model"	Part "workplace," part teamwork	Teamwork, personal leadership	Teamwork, personal leadership
Relationship of volunteer to governance	Clear differentiation between volunteer and paid staff	Somewhat clear differentiation between volunteer and paid staff	No paid staff, organization governed by member activists	Ambiguous difference between volunteer and paid staff

Source: Adapted from Rochester (1999).

Rochester (1999) describes the service delivery model as one in which most of the work of an organization is performed by volunteers who are recruited for specific duties, as for example in an agency providing social services. Paid staff members act in a supervisory capacity, in addition to recruiting and training volunteers. The relationship between paid staff and volunteers is clear and hierarchal. He also refers to this kind of volunteer utilization as the "workplace model" (p. 12). In this model volunteers are to be managed similarly to part-time, unpaid staff, a conception often found in the U.S. literature on volunteer involvement in nonprofit or government agencies (Brudney, 1990; Connors, 2011). In an organization that utilizes the support-role model, volunteers assist and supplement the work of paid staff members, for example, as aides and clerical personnel. The roles of volunteers and paid staff members are distinct, and volunteers participate in many aspects of the organization's operations. Volunteer management decisions are somewhat collaborative, and supervision may range from informal to structured, depending on the culture of the organization, the work to be done, and the requests of volunteers.

In the other two models in Rochester's framework, the distinctions and authority relationships between volunteers and paid staff members are more fluid. A distinctive feature of the co-worker model is the ambiguous nature of the relationship between volunteers and paid staff. In such organizations, paid staff and volunteers perform the same tasks and make decisions collaboratively (Rochester, 1999, p. 12); religious congregations, political parties, and professional associations offer examples (Gazley & Dignam, 2008). Management of volunteers is carried out by a "non-hierarchical team or collective" (Rochester, 1999, p. 17). Leaders of teams are most often, but not always, chosen from among the paid staff and should exhibit a leadership style described as "nurturing and enabling" (Rochester, 1999, p. 17). Similarly, in the member/activist model, all positions in the organization are held by volunteers who are bound by common commitment to a goal (Rochester,1999, p. 12). As in many day-care cooperatives, neighborhood associations, and self-help/mutual aid groups, volunteers must be governed by "peer-management" because hierarchal relationships do not exist (Borkman, 1999).

Meijs and colleagues (Meijs & Ten Hoorn, 2008; Meijs & Hoogstad, 2001) elaborate Rochester's (1999) framework in their contingency approach to volunteer management. They propose program management and membership management models, which are distinguished by the goal of the organization in which the volunteers donate their time, and the relationship between paid staff and volunteers (See Table 4). At one end of the organizational goal dimension lie campaigning (advocacy) and mutual support organizations, and at the other are service delivery organizations. At one extreme of the volunteer–paid staff dimension are volunteer-run organizations, which have no paid staff, and at the other are volunteer-supported organizations, in which volunteers assist paid staff in organizational operations. According to Meijs and colleagues (Meijs & Hoogstad, 2001; Meijs & Ten Hoorn, 2008), service delivery-type organizations, which typically are volunteer supported, require a program management style of management. In this setting, volunteer managers recruit volunteers for pre-identified tasks to meet organizational goals. By contrast, mutual support and campaigning organizations, which are dominated by volunteers, necessitate a membership management style. In this form, it is the responsibility of the volunteer manager to work collaboratively with volunteers to develop tasks that meet the social needs of the individual and the group (Meijs & Hoogstad, 2001; Meijs & Ten Hoorn, 2008). Table 4 provides a detailed description of the program management and membership management models.

Table 4 Program Management and Membership Management Model

Criteria	*Program Management*	*Membership Management*
Structure		
Flexibility of approach	From task to volunteer	From volunteer to task/assignment
Integration	Free-standing programs	Integrated approach
Direction of integration in national organization	Vertical	Horizontal (i.e., per branch)
Management	One single manager	Group of "managers"
Executive committee	Arm's length	Close by
Culture		
Organizational culture	Weak	Strong
Volunteer involvement	Low	High
Volunteer involvement in more than one organization	Often	Sometimes
Level of homogeneity among volunteers	Low	High
Relationships between volunteers	People do not know each other	People know each other well or very well
Volunteers' motivation 1	Goal orientated	Socially orientated
Volunteers' motivation 2	Increase in external status	Strengthening internal status
Process		
Cost of admission	Low social costs	High social costs
Cost of transfer	Low	High
Expectations	Explicit	Implicit
Recognition	On basis of performance	On basis of number of years as member
Hours spent/invested	Low	High
Environment		
Necessity of conforming to environment	Major	Minor
Possibility of conforming	Good	Poor

Source: Meijs & Hoogstad (2001).

Table 5 The Volunteer Involvement Framework

	Connection to Service	
Time for Service	*Affiliation Focus*	*Skill Focus*
Short-term Episodic	Strong planning and project-management skills, diplomacy, and passion	Flexibility and skills in recruitment and human resources
Long-term Ongoing	Knowledge of the organization's future direction, ample time to devote to volunteers, and strong interpersonal skills	Collaborative management style

Source: Adapted from Rehnborg (2009, p. 10).

Rehnborg (2009) presents another contingency approach to volunteer management based on the characteristics of the volunteer in her "Volunteer Involvement Framework." As illustrated in Table 5, Rehnborg's framework consists of two dimensions: a volunteer's connection to service and his or her "time for service" (Rehnborg, 2009, p. 10). Connection to service refers to whether the volunteer is affiliation focused (generalist) or skills focused (specialist). *Affiliation focused* refers either to a volunteer's motivation to become involved in a specific mission or to his or her desire to fulfill a requirement or goal of a group in which he or she is already involved. *Skills focused* refers to a volunteer who seeks to share his or her skills or one who seeks to gain skills through volunteer work. *Time for service* pertains to how much time a volunteer is willing to devote to service ("short-term" or "long-term") and whether the service is episodic or ongoing.

Cross-classification of Rehnborg's (2009) dimensions of the volunteer's connection to service and time for service results in four volunteer types or contingencies that confront social workers. "Short-term generalist" volunteers participate in such activities as corporate days of service, park clean-up events, or other time-intensive volunteer activities; "short-term specialist" volunteers provide donated services, for example, by psychologists and other health professionals or pro bono legal advice by attorneys on a one-time (or limited) basis; "long-term generalist" volunteers include mentors to youth, or unpaid individuals staffing homeless shelters; "long-term specialist" volunteers encompass volunteer firefighters, loaned executives, or professionals willing to make an ongoing commitment to social work agencies or clients (Rehnborg, 2009, p. 10).

Rehnborg (2009) associates these volunteer types with appropriate or recommended volunteer management techniques. For example, among the managerial traits needed to facilitate service by an affiliation-focused, short-term, episodic volunteer (short-term generalist) are strong planning and project-management skills, diplomacy, and passion. Managerial traits required to address the needs of an affiliation-focused, long-term, ongoing volunteer (long-term generalist) include knowledge of the organization's future direction, ample time to devote to volunteers, and strong interpersonal skills. In managing the skill-focused, short-term, episodic volunteer (short-term specialist), traits such as flexibility, as well as skills in recruitment and human resources, are important. Skills-focused, long-term, ongoing volunteers (long-term specialist) require a more collaborative management style.

1.5 Program/Organization-Focused

Macduff et al. (2009) propose a conditional model of volunteer management centering on volunteer program type. By contrast to the volunteer-based contingency models, this one proposes that the prevailing culture and worldview of the program or organization influence the appropriate management of volunteers. The authors incorporate as the dimensions of their framework the extent to which the worldview and culture of the volunteer program promote radical change versus regulation, and exhibit flexibility versus stability. With respect to the first dimension, a program embracing a high level of radical change (also called differentiation) seeks to influence dramatic shifts to existing structures, while programs with a primary focus on regulation (also called integration) strive to preserve the status quo. With respect to the second, the endpoints are flexibility (also referred to as discretion and subjectivity) and stability (also called control and objectivity). Highly flexible programs are distinguished by fluid environments and collaborative decision making, and those characterized by a high degree of stability have established protocol and ordered, systematic programming.

As shown in Figure 1, four categories of volunteer programs derive from cross- classifying the worldview and culture of the volunteer program as promoting radical change versus regulation, and exhibiting flexibility versus stability. "Traditional" volunteer programs focus on regulation and value stability over flexibility, as, for example, in most government social service agencies. This type of program is hierarchal, and volunteer coordinators should rely on standard or best practices in their management of volunteers. "Social-change" programs are characterized by an emphasis on radical change and stability. The mission of this program type is the highly ambitious "transformation of societies, systems, programs, and services based on perceptions of unmet needs of various population groups" (Macduff et al., 2009, p. 411), and thus volunteer managers should be prepared to organize and lead activist volunteers. A "serendipitous" program features a high degree of flexibility and regulation. This model has minimal volunteer structure and is more often "coordinated than managed" (Macduff et al., 2009, p. 413). Serendipitous volunteers often give service informally or spontaneously, and management must support these individuals with participative decision making and collaboratively designed volunteer duties.

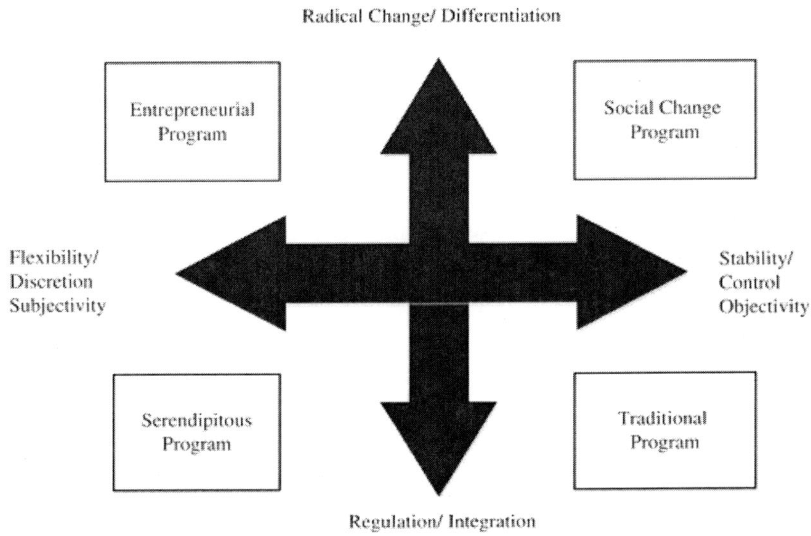

Figure 1 Ideal Program Types Based on Organizational Culture and Worldview.

The final program type proposed by Macduff et al. (2009) is the entrepreneurial program. Given its great flexibility and interest in radical change, the worldview and culture of an entrepreneurial program is opposite to a traditional program; an entrepreneurial program may consist of a single individual committed to addressing societal ills through empowerment solutions. Although an entrepreneurial volunteer "will not be managed" (Macduff et al., 2009, p. 415), he or she might become affiliated with an established program when granted autonomy and a chance to make a difference.

2. DISCUSSION: APPLYING PRINCIPLES TO PRACTICE

We propose that the various approaches to volunteer administration and management that we have delineated can both co-exist and inform social workers and managers in human service organizations. The most basic distinction that these professionals need to make in working with volunteers is whether the tasks assigned to volunteers are the same as, or different than, those performed by paid staff in the organization. As Handy, Mook, and Quarter (2008) and their colleagues (Chum, Mook, Handy, Schugurensky, & Quarter, 2013) observe, some interchangeability between the work performed by paid staff and volunteers is endemic to work life in many organizations and is to be expected. To the extent that volunteers perform the same tasks as paid staff, the universalistic practices, which are rooted in the metaphor of the workplace, would seem to apply, including job descriptions, written policies and procedures, and orientation and supervision, among others (see Tables 1 and 2). Likewise, a hierarchical element usually applies, as, for example, in Rochester's (1999) service delivery or workplace model of volunteer management (Table 3). Because volunteers largely reflect paid workers, except that they are not compensated monetarily, they can be managed similarly to part-time staff in these situations.

As the contingency approaches demonstrate, however, these situations are not the only ones in which social workers and managers in human service organizations work with volunteers. And it is in these more fluid circumstances that the conditional approaches to volunteer administration are better suited to the volunteer management task and more effective than the universalistic approach. When volunteers do not simply replicate the work of paid staff, the conditional approaches elaborated earlier are more appropriate and helpful. Two sources of fluidity are apparent and govern the choice of a fitting (contingency) volunteer management model.

First, in some volunteer programs, volunteer involvement is based on the contribution of sporadic and oftentimes "micro" bits of talent and energy (as little as an hour or two) in so-called episodic volunteer programs or opportunities. Although the contribution of hours and skills donated in such programs may be highly useful in the aggregate to the organizations that receive them, emulating a paid work "schedule" and management practices under these circumstances is impractical and more wasteful than efficient. In these situations social workers and human service managers should give serious consideration to the conditional approach to volunteer management explicated by Rehnborg (2009), who bases her management prescriptions according to whether volunteer involvement is short-term and episodic versus long-term and ongoing. Rehnborg's framework offers useful advice for the management of episodic volunteers (or long-term volunteers), depending on whether the goals of the participants are affiliation focused or skill focused (Table 5). Similarly, in their framework, Macduff et al. (2009) present a serendipitous model of volunteer management grounded in the sporadic and sometimes spontaneous involvement of volunteers (Figure 1). In organizations that feature this kind of episodic volunteer involvement, the management practices recommended by Macduff et al. (2009) are far more fitting and appropriate than the workplace analogue.

Second, in many organizations the authority relationships between volunteers and paid staff are more fluid than assumed in the universalistic approaches. Absent a hierarchical relationship governing paid staff and volunteers, the universalistic approaches fail to give relevant and useful guidance. Here, social workers and managers in human service organizations should turn again to the conditional approaches for appropriate recommendations. Macduff et al. (2009) describe the active involvement of volunteers in highly flexible, open-ended programs intended to affect systematic social change. Consider, for example, the task of working with advocacy groups of volunteers in the community, committed to improving the lot of new immigrants, or promoting the sustainability of the environment, or arresting social inequities. These circumstances do not permit volunteer "management," and the universalist approach will not do. Instead, Macduff et al. (2009) present more constructive advice for working with volunteers in this situation in their entrepreneurial volunteer model and social change model.

Social workers encounter other, more prosaic situations in which authority is elusive. For example, the contingency models support social workers who are involved in the management of voluntary associations, collaboratives, self-help groups, and the like. Meijs and colleagues (Meijs & Hoogstad, 2001; Meijs & Ten Hoorn, 2008) present extended recommendations in the membership management model, as does Rochester (1999) in the member/activist model. Rather than directing clients, former clients, members, and lay citizens in these group-based efforts, social workers need to embrace a more collaborative style of management.

3. CONCLUSION: SOCIAL WORKERS AND PROFESSIONALISM IN VOLUNTEER MANAGEMENT

Our analysis reveals universal and contingency approaches to the volunteer management task. The response of universalist researchers and practitioners to issues of management is most often a behavioral prescription in the form of a generic framework of best practices. Universal best practices are "one size fits all" management solutions, meant to apply to all organizations and volunteers regardless of mission, organizational culture, and volunteer characteristics. Despite the breadth of application assumed by the universalists, however, the approach seems best suited to those volunteer programs in which volunteers mainly replicate the jobs and roles of paid staff, and the relationship between the two parties remains hierarchical. Given the enormous variety of social work and human services jobs, programs, and organizations, as well as the diversity of the volunteers involved, this approach seems narrow, if not simplistic.

Researchers and practitioners who propose a conditional approach to volunteer management help to clarify the highly diverse range of volunteer involvement in social work practice. In addition to acknowledging a place for the workplace model under the appropriate circumstances, their frameworks provide useful guidance when this model breaks down with respect to the fluidity of the types of roles assumed by volunteers and/or their authority relationships with paid staff. Thus, taking a contingency approach does not mean abandoning the suggestions of the universalists. Instead, our analysis shows that the universalist approach is subsumed in the contingency models as valid and useful, but only under certain conditions, for example, in Rochester's (1999) service delivery model, in the Macduff et al. (2009) traditional program, and in the Meijs and colleagues (Meijs & Hoogstad, 2001; Meijs & Ten Hoorn, 2008) program management model. Researchers who take a conditional approach to volunteer administration consider factors pertaining to a program/organization or its volunteers—or both—in determining recommendations for management. Inclusion of particular factors in making management decisions results in a more tailored approach.

With the growing pressure on them to involve volunteers, social workers need to develop and improve their skills as volunteer administrators, while continuing to provide for client welfare. We believe that involving volunteers effectively begins with understanding the "universals" of volunteer management and applying them where and when appropriate. However, becoming a professional volunteer administrator means, in part, recognizing differences in program/organizational context and in volunteers that may affect the volunteer management task, and then having the expertise and ability to implement appropriate techniques. By presenting a wide array of contingencies of volunteer management systems, this article assists social workers in contextualizing their existing knowledge on collaborating with volunteers, thus helping them to become professionals capable of working with volunteers in changing organizational circumstances (Leicht & Fennell, 1997). In the future, as in the past, these skills will be crucial for social workers committed to enacting real social change.

ACKNOWLEDGMENT

The authors thank Allison R. Russell for her assistance with this article. We thank Vanessa Lacer for assistance with an earlier draft.

REFERENCES

1 Becker, D. G. (1964). Exit Lady Bountiful: The volunteer and the professional social worker. Social Service Review, 38(1), 57–72.

2 Borkman, T. J. (1999). Understanding self-help/mutual-aid: Experiential learning in the commons. New Brunswick, NJ: Rutgers University Press.

3 Brudney J. L. (1990). Fostering volunteer programs in the public sector: Planning, initiating, and managing voluntary activities. San Francisco, CA: Jossey-Bass.

4 Brudney, J. L., & Meijs, L.C.P.M. (2009). It ain't natural: Toward a new (natural) resource conceptualization for volunteer

management. Nonprofit and Voluntary Sector Quarterly, 38(4), 564–581.

5 Chum, A., Mook, L., Handy, F., Schugurensky, D., & Quarter, J. (2013). Degree and direction of paid employee/volunteer interchange in nonprofit organizations. Nonprofit Management and Leadership, 23(4), 409–426.

6 Connors, T. D. (2011). The volunteer management handbook: Leadership strategies for success (2nd ed.). New York, NY: Wiley.

7 Cuskelly, G., Taylor, T., Hoye, R., & Darcy, S. (2006). Volunteer management practices and volunteer retention: A human resource management approach. Sport Management Review, 9, 141–163.

8 Delery, J. E., & Doty, D. H. (1996). Modes of theorizing in strategic human resource management: Tests of universalistic, contingency, and configurational performance predictions. Academy of Management Journal, 39(4), 802–835.

9 Ellis, S. J. (2010). From the top down: The executive role in successful volunteer involvement (3rd ed.). Philadelphia, PA: Energize.

10 Gazley, B., & Dignam, M. (2008). The decision to volunteer: Why people give their time and how you can engage them. Washington, DC: ASAE and the Center for Association Leadership.

11 Hager, M. A., & Brudney, J. L. (2011). Universalistic versus contingent adoption of volunteer management practices. Paper presented at the Annual Meeting of the Association for Research on Nonprofit Organizations and Voluntary Action (ARNOVA), Toronto, ON, Canada, November 16–19.

12 Handy, F., Mook, L., & Quarter, J. (2008). The interchangeability of paid staff and volunteers in nonprofit organizations.Nonprofit and Voluntary Sector Quarterly, 37(1), 76–92.

13 Investing in Volunteers. (2010). U.K. quality standard for organisations that involve volunteers. Retrieved fromhttp://iiv.investinginvolunteers.org.uk/images/stories/theiivstandard2010.pdf

14 Leicht, K. T., & Fennell, M. L. (1997). The changing organizational context of professional work. Annual Review of Sociology,23, 215–231.

15 Macduff, N., Netting F. E., & O'Connor, M. K. (2009). Multiple ways of coordinating volunteers with differing styles of service. Journal of Community Practice, 17, 400–423. [Taylor & Francis Online]

16 Machin, J., & Paine, A. E. (2008). Management matters: A national survey of volunteer management capacity. London, UK: Institute for Volunteering Research. Retrieved from http://www.ivr.org.uk/ component/ivr/management-matters

17 McCurley, S., & Lynch, R. (2011). Volunteer management: Mobilizing all the resources of the community (3rd ed.). Philadelphia, PA: Energize.

18 Meijs, L., & Hoogstad, E. (2001). New ways of managing volunteers: Combining membership management and programme management. Voluntary Action, 3(3), 41–61.

19 Meijs, L., & Ten Hoorn E. (2008). No "one best" volunteer management and organizing: Two fundamentally different approaches. In M. Liao-Troth (Ed.), Challenges in volunteer management (pp. 29–50). Charlotte, NC: Information Age.

20 Netting, F. E., Nelson, H. W., Borders, K., & Huber, R. (2004). Volunteer and paid staff relationships: Implications for social work administration. Administration and Social Work, 28(3–4), 69–89.

21 Paull, M. (2002). Reframing volunteer management: A view from the West. Australian Journal on Volunteering, 7, 21–27.

22 Perlmutter, F. D., & Cnaan, R. A. (1994). Challenging human service organizations to redefine volunteer roles.Administration in Social Work, 17(4), 77–95. [Taylor & Francis Online]

23 Rehnborg, S. J. (2009). Strategic volunteer engagement: A guide for nonprofit and public sector leaders. Austin, TX: University of Texas, RGK Center for Philanthropy & Community Service. Retrieved fromhttp://www.volunteeralive.org/ docs/Strategic%20Volunteer%20Engagement.pdf

24 Rochester, C. (1999). One size does not fit all: Four models of involving volunteers in voluntary organizations. Voluntary Action, 1(2), 47–59.

25 Safrit, R. D. (2010). Evaluation and outcome measurement. In K. Seel (Ed.), Volunteer administration: Professional practice (pp. 313–371). Markham, Ontario, Canada: LexisNexis Canada.

26 Schwartz, F. (1984). Voluntarism and social work practice: A growing collaboration. Lanham, MD: Rowman and Littlefield.

27 Sherr, M. L. (2008). Social work with volunteers. Chicago: Lyceum Books.

28 Studer, S., & von Schnurbein, G. (2013). Organizational factors affecting volunteers: A literature review on volunteer coordination. Voluntas, 24, 403–440.

29 Theilen, G.L., & Poole, D.L. (1986). Educating leadership for affecting community change through voluntary associations. Journal of Social Work Education, 22(2), 19–29.

30 UPS Foundation (2002). A guide to investing in volunteer management: Improve your philanthropic portfolio. Points of Light Foundation and Volunteer Center National Network. Retrieved fromhttp://www.pointsoflight.org/pdfs/invest_vm_guide.pdf

31 Vinton, L. (2012). Professional administration of volunteer programs now more than ever: A case example. Administration in Social Work, 36(2), 133–148.

32 Wilson, J. (2012). Volunteerism research: A review essay. Nonprofit and Voluntary Sector Quarterly, 41(2), 176–212.

CHAPTER 9

The Mega-Event Syndrome: Why So Much Goes Wrong in Mega-Event Planning and What to Do About It

Martin Müller[a,b*]

[a] University of Zurich

[b] University of Birmingham

ABSTRACT

Problem, research strategy, and findings: Mega-events such as the Olympic Games and the Football World Cup have become complex and transformative undertakings over the last 30 years, with costs often exceeding USD $10 billion. These events are currently planned and governed in ways that produce adverse effects for cities, regions, and residents. This study identifies a *mega-event syndrome*, a group of symptoms that occur together and afflict mega-event planning, including overpromising benefits, underestimating costs, rewriting urban planning priorities to fit the event, using public resources for private interest, and suspending the regular rule of law. I describe each of these symptoms, providing empirical examples from different countries and mega-events, examining the underlying causes. The research is based on material from field visits to mega-event sites in 11 countries as well as 51 interviews with planners, managers, politicians, and consultants involved in mega-event planning.

Takeaway for practice: To curb the mega-event syndrome, I propose both radical and incremental policy suggestions. The most crucial radical change that an event host could make is to not tie mega-events to large-scale urban development, avoiding higher risks that create cost overruns, substandard construction quality, and oversized infrastructure not suitable for post-event demands. Further, event hosts should bargain with event-governing bodies for better conditions, earmark and cap public sector contributions, and seek independent advice on the costs and benefits of mega-events. Event-governing bodies, for their part, should reduce the size and requirements of the events.

KEYWORDS

mega-events, impacts, sport, Olympic Games, Football World Cup

Mega-events are one-time occasions of a fixed duration that attract a large number of visitors and have worldwide reach. They come with significant costs and long-term impacts on the built environment and the population of the host countries or cities. The Olympic Games and the Football (Soccer) World Cup are the largest mega-events, but events such as World's Fairs (Expos), regional games (e.g., Pan

American Games, Asian Games), and single sports events such as the Super Bowl are also mega-events.

In the past three decades, mega-events have reached a size that has made them transformative ventures for entire cities, regions, and sometimes whole countries. Such events now routinely command more than USD $10 billion in capital investment, take up several hundred acres of land, and require the infrastructure to accommodate, move, and keep secure hundreds of thousands of visitors and tens of thousands of athletes, officials, and journalists. Planning for mega-events appropriates resources, monopolizes public attention, can suspend the normal rule of law, and often rewrites urban and regional development plans. What were once primarily sports events or exhibitions have become urban events: occasions for large-scale urban transformation (Essex & Chalkley, 1998; J. R. Gold & Gold, 2008; Hiller, 2006; Kassens-Noor, 2012).

There is considerable disagreement about the nature of mega-event impacts. Cities and countries bidding to host the events often see them as boons to urban development, a view endorsed by event-governing bodies such as the International Olympic Committee (IOC) and some consulting companies and development agencies (German Development Agency [GIZ], 2013; PricewaterhouseCoopers, 2011). Some scholars, too, have claimed that mega-events are catalysts for urban development and "accelerate [a city's] infrastructural development by up to 10 years" (Preuss, 2004, p. 232). A few studies suggest that mega-events can create an economic benefit to host cities (Gratton, Shibli, & Coleman, 2005), and that cities and countries can leverage them to improve their image and empower community action (Chalip, 2006; Grix, 2012). A focus on planning for the legacies of mega-events—what is left after the event—has served to reinforce the idea that mega-events can be positive forces for urban development (Holt & Ruta, 2015; Jago, Dwyer, Lipman, van Lill, & Vorster, 2012).

At the same time, the negative impacts of mega-events on cities and regions are well documented and occur in almost every case. Cost overruns, schedule slips, oversized infrastructure, and social polarization dash the high expectations for positive urban development resulting from such events (e.g., Boykoff, 2014; Cottle, 2011; Gaffney, 2010; Hayes & Horne, 2011; Horne,2007; Shin & Li, 2013). Thus, since 1960, without exception, the Olympic Games have gone over budget, on average by 179% (Flyvbjerg & Stewart, 2012). After the 1994 World Cup in the United States, host cities experienced a net economic loss rather than the predicted gain (Baade & Matheson, 2004). The Olympic Summer Games in Athens cost at least 3.4% of the gross domestic product (GDP) of Greece at the time and left a legacy of underused sports facilities and environmental destruction (M. M. Gold, 2007). In Rio de Janeiro, preparations for the 2014 Football World Cup and the 2016 Olympic Games exacerbated sociospatial polarization, as authorities evicted and resettled tens of thousands of residents (de Paula, 2014).

In this study, I suggest that there are common dynamics that plague mega-event planning everywhere in the world to a greater or lesser degree, or what I call the *mega-event syndrome*. The concept of "syndrome," drawn from medicine, suggests the presence of a set of symptoms that occur together, recognizing that these symptoms may be interrelated and have shared underlying causes (Schellnhuber et al., 1997). On the basis of 51 interviews with mega-event planners, managers, politicians, and consultants; field visits to mega-event sites in 11 countries; and official documents, polls, and media reports, I identify seven major symptoms that together form the mega-event syndrome and are shown in Table 1: overpromising benefits, underestimating costs, event takeover, public risk taking, rule of exception, elite capture, and event fix. Investigating these symptoms together allows us to see them as the result of a shared complex of problems, shifting the focus toward the causes and drivers of negative outcomes. In so doing, I propose both radical and incremental policy suggestions to reduce the prevalence and size of the mega-event syndrome for future hosts.

Table 1. The mega-event syndrome: symptoms and consequences.

Symptom	Description	Consequences
1. Overpromising of benefits	Overestimating positive effects of mega-events	• Misallocation of resources • Loss of trust with citizenry
2. Underestimation of costs	Actual budget > planned budget	• Misallocation of resources • Profiteering • Subpar construction quality • Budget shortfalls
3. Event takeover	Event priorities become planning priorities	• Event needs displace urban infrastructure needs • Oversized infrastructure • Unfinished infrastructure
4. Public risk taking	Public takes risk for private benefits	• Public funds for limited or no public benefits • Profiteering
5. Rule of exception	Suspension of regular rule of law	• Displacement • Reduced public oversight • Limited public participation
6. Elite capture	Inequitable distribution of resources	• Spatially uneven urban landscape • Gentrification
7. Event fix	Mega-events become seemingly quick fixes for major planning challenges	• Event determines national priority for funding • Bypassing of regular planning process • Waste of resources on event as lever for urban development

1. RESEARCH APPROACH AND MATERIALS

The primary empirical basis for this study rests on 51 semistructured interviews and material from field visits to mega-event sites in Brazil (Rio de Janeiro), China (Beijing), Canada (Calgary, Vancouver), Germany (Berlin, Munich), Italy (Torino), Korea (Gwangju, Seoul, Yeosu), Russia (Kazan, Moscow, Sochi), Spain (Sevilla), Ukraine (Lviv), the United Kingdom (London), and the United States (Salt Lake City [UT]). I used site visits to collect firsthand material on the planning and afteruse of mega-event - facilities in a range of countries to avoid a national bias in the analysis.

I conducted the interviews between 2010 and 2014 with staff from event-governing bodies (the IOC and FIFA [Fédération Internationale de Football Association, the governing body for the Football World Cup]), organizing committees, consultants to cities and -organizers, politicians, and planners. I selected interviewees who had a strategic, not just an operational, role in the planning and preparation of mega-events or in managing the infrastructure after the event across a number of functional areas such as transport, environment, venue planning, accommodation planning, and knowledge management. I avoided a potential sampling bias that could have occurred when focusing on just one area of mega-event planning. All potential interviewees, except two, were available and willing to speak with me. In those two cases, I contacted people with similar functions.

The interviews covered the following thematic areas: the rationale for bidding and the process of assembling the bid documents; the guiding principles and priorities of planning for and managing the event and the period afterwards; challenges, tradeoffs, and unexpected events during the planning process; the influence of different stakeholders on the planning and management process and on outcomes; positive and negative outcomes; and the reasons why some outcomes differed from initial expectations. The exact questions varied according to the role of the person interviewed and the specific mega-event. I used an interview guide for conducting interviews that -contained open-ended questions reflecting these areas of inquiry. While I did not share the exact questions in -advance, I did outline the general theme of the interview in all my requests.

I took several steps, common for qualitative methods (see McMillan & Schumacher, 1997), to establish that the material collected was valid and that I correctly understood the responses made by those I interviewed. First, I made sure that interviewees were comfortable with the interview language. Most interviewees were fluent in

English, so I conducted most interviews in -English. In one case the interviewee preferred Russian, and in one case the interviewee preferred German, both languages I speak fluently. Second, I audio-recorded all interviews, after receiving permission, and transcribed them in the original language to allow for exact -reproduction.

After each of the first five interviews, I revised the guide where I thought interviewees had not understood the intent of the questions or the wording of questions was imprecise. I also asked interviewees if they agreed with statements made by other interviewees to determine the level of agreement. I then coded interviews into themes and subthemes to compare the material across interviewees and field sites. I did not work with a preestablished -codebook, but generated codes from the material to reduce the risk of preconceived ideas influencing how I structured the material.

When I quote material here, I rely on experiences that interviewees reported more than once to avoid basing the analysis on unusual statements. During the site visits, I was able to compare the statements of those interviewed against my own observations.

I believe this method is a conservative approach to understanding the critical aspects of mega-event planning because I interviewed people who may have an interest in downplaying the problems and the challenges of mega-events in which they themselves were involved. My -approach guards against an excessively pessimistic stance that could result from interviewing those opposed to mega-events. Moreover, these interviews and my -interpretation of the responses are the result of five years of fieldwork on various mega-events.

In addition to interviews and site visits, I use material such as bidding documents and hosting guidelines, surveys, and media reports, which are publicly available. To -complement this primary material, I integrate additional evidence from the existing academic literature on mega-events. These materials form the basis for extracting the seven symptoms of the mega-event syndrome, which I describe below.

2. THE MEGA-EVENT SYNDROME

2.1 Symptom 1: Overpromising of Benefits

This symptom refers to the mismatch between the expected and the actual benefits of a mega-event. This is particularly evident in the bid book, the document organizers submit to an event-governing body such as the IOC or FIFA as an application to host the event. The bid book makes promises about the outcomes of the event and becomes part of the contract between a host and the event-governing body. Bid books promulgate a vision of urban transformation, with associated costs and benefits. They serve as key documents for decision-makers to evaluate whether to host an event and fund the expenditure.

Yet, the promises in bid books are not a realistic -statement of what cities want to do and can achieve with the event-governing bodies. As one member of staff in an organizing committee stated:

The bid book is science fiction—an imaginary case.... International consultants put into the bid book what they knew that the IOC would like and buy, and when we got the Games it was all of a sudden like, "Whoooa, we have to deliver on all that and we had no idea how to do that." (interview with staff of organizing committee)

Overpromising is also widespread when prospective hosts forecast the economic impacts of mega-events. The mayor of Chicago, for example, in the city's bid to host the 2015 Olympic Games, promised that hosting the event would create 315,000 new jobs, a forecast more than four times the estimate for the Olympics in Atlanta (GA) in 1996 and dismissed as crazy by sports economist Victor Matheson (quoted in Pletz, 2009; see also Eisinger, 2000; Matheson, 2008). However unrealistic a given figure for economic impacts may be, it still produces an important psychological anchoring effect, creating the impression that mega-events come with economic gains (Tversky & -Kahneman, 1974). Thus, in a survey in 16 countries, 68% of respondents reported that hosting the Olympic Games is an "opportunity for economic development" and 73% thought it "leaves the host city with many benefits" (Kantar, 2014).

Sports economists, however, are in rare unanimity that providing subsidies for mega-events is, at best, a suboptimal use of public money and incurs significant opportunity costs (Baumann & Matheson, 2013; Crompton, 1995; Jennings, 2012; Madden, 2006; Matheson, 2008; Mills & Rosentraub, 2013; Zimbalist, 2015). The 1994 Winter Olympic Games in Lillehammer, for example, cost 100 times more than a tourism development program the Norwegian government ran in parallel, yet the increase in overnight stays due to the Games was only twice as large as that of the tourism program (Teigland, 1999).

Despite this poor record, cities and countries continue to bid for mega-events because the event promoters tend to profit most from such events. Mega-events bring prestige for sports officials; profit for property developers, construction firms, and landowners; and a chance to shine for politicians. It is this unequal distribution of costs and benefits (see Logan & Molotch,1987) that continues to produce bids, although the promised benefits hardly ever materialize.

2.2 Symptom 2: Underestimation of Costs

This symptom is the alter ego of overpromising benefits. Like most mega-projects (Flyvbjerg, Holm, & Buhl, 2002; Priemus, Flyvbjerg, & van Wee, 2008), mega-events overrun their budget, sometimes massively so. The most blatant example is the 2010 Commonwealth Games in New Delhi (India); initially estimated to cost about USD $50 million, the actual total budget was more than USD $4 billion (INR 280 billion; High Level Committee for -Commonwealth Games 2010, 2011). The Olympic Games always overrun their initial budget. The average cost -overrun for the Olympic Games since 1960, including operating and direct capital costs, is 179%: much more than the 27% average overrun for transport mega-projects (Flyvbjerg & Stewart,2012).

Many factors drive organizers to underestimate the costs of mega-events. First, unlike other mega-projects, mega-events have a *fixed deadline*. Postponing the opening date is not an option when preparations fall behind -schedule, so event organizers will hire additional workers or add nightshifts. In addition, when construction has to be rushed, competitive bidding (tendering) rules are relaxed, reducing competition and driving prices up further.

Second, mega-events cannot be gradually ramped up; they are *nondivisible* (i.e., all competitions need to run from day one at full force). Thus, there is no room for trial and error, and that comes with a high cost. Third, when work is running late, contractors engage in *profiteering;* they know that event organizers are dependent on them to finish work in time. The closer the opening date looms, the more contractors will ask for large premiums to finish work in time. Fourth, mega-event planners operate with *large contingencies,* lacking knowledge about the demand -patterns during previous events in other cities. They build infrastructure larger than needed because there is substantial uncertainty about the demand, such as for public transport on a particular day during the event. This -uncertainty results from the exceptional character of the event and is exacerbated by lack of knowledge of usage at previous events.

Fifth, mega-events have a *long implementation period,* often more than 10 years from the initial idea to the final execution. During this time, many of the initial assumptions are subject to change. Inflation, for example, can increase due to the additional demand that a mega-event creates. External events, such as terrorist attacks, can increase costs for security. New regulations or demands can make it necessary to expand the scope of the budget. What compounds this uncertainty is that the event-governing bodies tend to leave some requirements deliberately vague in their contracts with host cities, postponing concrete specifications to a later stage, as in the FIFA Host City Agreement. This makes it difficult for cities to engage in reliable advance planning and budgeting and increases the risk of changes at short notice that increase costs.

Sixth and last, there is an incentive for event promoters to *misrepresent the true cost* of an event. There are two main reasons for this. Bids need to garner public support, both because governing bodies of the events demand that a majority of the population be in favor of the event and because in many countries the public or the government can—but do not have to—initiate referenda on bids. Munich, St. Moritz, and Krakow, for instance, had to withdraw from bidding for the 2022 Winter Games because of failed referenda. Also, potential host cities compete at the national level for nomination as the national candidate for the event, often with guarantees of public subsidies. It is imperative that a proposed project looks as inexpensive as possible to engage public support and improve chances for nomination as the national candidate with the attached national subsidies.

2.3 Symptom 3: Event Takeover

Mega-event priorities often displace long-term urban development priorities. Instead of the event becoming an instrument for urban development, urban development becomes the instrument for the event. The most extreme example is perhaps Rio de Janeiro, which in the space of 10 years hosted the Pan American Games (2007), the Football World Cup (2014), and the Olympic Games (2016), ending up "with an urban structure that was—for over two decades—purely driven by sporting events" (Kassens-Noor, 2012, p. 105). The mayor of Rio, Eduardo Paes, acknowledged that mega-event priorities are what drive Rio's development planning: "The Olympic plan is Rio's plan, and Rio's plan is the Olympics" (quoted in Nuzman, 2011).

Event takeover has two major characteristics. The first is that event requirements crowd out actual urban infrastructure needs, both in spatial and financial terms. Venues for hosting an event occupy space, often in prime locations of a city, that could be used for other public facilities with broader or more sustained public benefit. The venues for the Summer Olympics, for example, require an area of almost 700 hectares (1,730 acres), not including ancillary infrastructure such as the Olympic Village, ceremonial spaces, parking areas, etc. (Long, 2013). These requirements increase competition for the use of scarce urban land in densely populated cities.

But crowding out also happens in a financial sense: As the event nears and budgets are squeezed, sports-related facilities prevail over wider urban improvements. Stadia, athlete accommodation, media centers, airport extensions—all of them contractual obligations to the governing bodies of the event—appropriate budgets originally meant to improve housing, community facilities, or public transport, undermining long-term urban development priorities and public preferences.

Second, building for the demands of the event often does not reflect post-event needs. Fulfilling the requirements of event-governing bodies for airport or public transit capacity, for example, requires the host city to tailor the size of the project to peak event demand, which is often too large for efficient post-event use. Since event-governing bodies set the requirements for infrastructure, but do not pay for its delivery, they have an incentive to demand excessively large stadia, airports, or hotel capacities. The airport in Lviv (Ukraine), upgraded to meet the requirements of UEFA, the governing body of the European Football Championship, now boasts a capacity of 20,000 passengers per day. However, it was running at 10% of that capacity in 2013, one year after Ukraine hosted the championship.

Building many event venues in the same area exacerbates the problem of oversized infrastructure. This risk is particularly high for Olympic parks, which have several stadia with large seating capacities. To accommodate the transit demand during the event, organizers build large capacities and plan rail transit lines to optimize the movement of event visitors. However, these arrangements often do not reflect the post-event demands. After the 2014 Winter Games in Sochi (Russia), for example, the combined road and rail link between the airport and the venue -clusters, built for more than USD $10 billion, became too expensive to operate. Sydney (Australia), too, built a transit system for the 2000 Summer Games that was too large for post-event use; moreover, its route system is not geared to post-event travel patterns (Kassens-Noor, 2012).

2.4 Symptom 4: Public Risk Taking

The public sector provides the deficit guarantee demanded by event-governing bodies in almost all cases, thus accepting the risk of failure. The government must deliver the defined facilities and infrastructure and cover all budget shortfalls, while the profits accrue to the event-governing bodies and the contractors. This guarantee encourages profiteering: Contractors know that the government is obliged to finish the project regardless of price.

Cities hosting mega-events portray them as a way of attracting private capital in an environment of increasing interurban competition (Andranovich, Burbank, & Heying, 2001; Burbank, Andranovich, & Heying, 2002; French & Disher, 1997; Hall,2006). But investors tend to be wary of investing in sports facilities and ancillary infrastructure (Long, 2012; see Rosentraub [1999] for the case of sports facilities in the United States). The taxpayers almost always pay a higher share for the event than planned. The organizers of the 2014 Football World Cup in Brazil and the 2014 Winter Games in Russia expected private investment to make up most of the funding for the event, but in both cases the public was forced to pay more than 95% of the costs (Fund for the Fight Against Corruption, 2014; -Gaffney, 2014). London and Vancouver tried to attract private investors for the Olympic villages, but had to bail them out after the financial crisis (Scherer, 2011).

2.5 Symptom 5: Rule of Exception

Mega-events are exceptional happenings and so, too, is the time during which the hosts prepare for them. The preparation for most mega-events includes special legislation that suspends the regular rules to facilitate hosting the event (Coaffee, 2014; S·nchez & Broudehoux, 2013). Many governments pass laws that introduce exceptions in areas such as taxation, immigration, property rights, urban planning, and freedom of speech. Governments, for -example, have to provide tax exemptions on some revenues accruing to event-governing bodies and award franchises or monopolies that go beyond existing legislation (Louw, 2012). For the World Cup in Brazil, tax authorities -estimated that these exemptions would cost roughly USD $250 million (BRL 559 million) in foregone tax revenue (de Paula, 2014).

But cities and countries may also curtail citizen rights without explicit requirements from event-governing bodies. Vancouver, for example, introduced a bylaw outlawing placards, posters, and banners that did not celebrate the Olympic Games in 2010 (Boykoff, 2011). The mayor of Boston signed a decree that forbade city employees from criticizing the Olympics during the city's bid to host the event in 2024 (Levenson, 2015). Both cities, however, had to repeal these decisions following legal challenges.

It is common for hosts to expropriate private property and displace residents during the run-up for a mega-event in democratic and authoritarian countries alike (Centre on Housing Rights and Evictions [COHRE], 2007; Porter et al., 2009). Targeted legislation facilitates expropriation and displacement (Davis, 2011; Olds, 1998). A special report by COHRE, a nongovernmental organization (NGO) supporting the cause of equal housing rights, states that several hundred to several thousand people were displaced for a number of mega-events, mostly to clear the ground to build facilities to host the event (COHRE, 2007).

2.6 Symptom 6: Elite Capture

Proponents often justify hosting a mega-event because it will create broad societal benefits. Yet, mega-event -planning tends to privilege local business and real estate interests, global corporations, and the cronies of the political elites (Andranovich et al., 2001; Shaw, 2012; compare with Logan & Molotch, 1987). In Brazil, for example, several well-connected construction firms—known as the "four sisters"—pocketed the lion's share of public contracts for the 2014 World Cup, producing vast cost overruns (Belis·rio, 2014). Moreover, new infrastructure often gives priority to airports, which cater to the highly mobile strata of society. Even the sports stadia themselves have a -tendency to turn into socially exclusive spaces if ticket prices multiply to recoup at least part of the high cost of the venues, as in Brazil (Gaffney, 2014).

Event-induced gentrification contributes to elite capture and is a phenomenon that has become a familiar sight in most mega-event host cities that harness events for urban regeneration, from Atlanta (Rutheiser, 1997) and Sydney to Vancouver (Lenskyj, 2008), London (Watt, 2013), and Rio (Gaffney, 2010). In Stratford in East London, where the Olympic Park is located, the Olympic Games accelerated gentrification and displacement of residents (Watt, 2013). Mega-events thus generate both the financial capital required for urban regeneration and the symbolic capital needed to make areas attractive for wealthier residents.

At the same time, elite capture curtails public oversight and participation. While event organizers want public support during the bid, the participation of the public is often considered expendable or is reduced once a city has won. Planning for mega-events then turns into a technocratic process of delivery and democratic demands become risks that threaten to delay the planning and construction process of the event (Raco, 2014; see also Andranovich et al., 2001; Hiller, 2000). Event-governing bodies sometimes perceive even democratic decision making as a nuisance, delaying and derailing the preparation for the event. JÈrÙme Valcke, Secretary General of FIFA, underscored that by saying, "Less democracy is sometimes better for organizing a World Cup" (quoted in Reuters, 2013).

2.7 Symptom 7: Event Fix

Event fix occurs when mega-events become seemingly quick fixes to fast-track major urban development projects: a "shot of adrenalin" for cities to clean up their act and get projects done that would otherwise have stalled or never happened. As the deputy mayor of Lviv, host to the 2012 European Football Championship, said:

Hosting a mega-event is like when the aunt from Canada comes to visit—you repair the house, you make a spring cleaning—when over the whole year you keep throwing around the rubbish. (podium discussion, November 2012)

As a result, cities and countries spend large sums on hosting an event to justify or create enough pressure to attain the non-sports investments they want to pursue. The rule of exception provides support for doing this: For example, where proper environmental impact asssessments, full tenders, or due diligence are considered too cumbersome, organizing a mega-event permits the short-circuiting of regular planning procedures.

Mega-events produce or force consensus where planning proposals may not have been able to do so, thus pushing through potentially contentious projects as necessities for the event. Mega-events also can serve as a lever to extract funds for local governments from the central government. As the former mayor of London, Ken Livingstone, said:

I didn't bid for the Olympics because I wanted three weeks of sport. I bid for the Olympics because it's the only way to get the billions of pounds out of the Government to develop the East End. (quoted in Davies, 2008)

As a consequence, the central government allocates funding for urban development, not based on proven need or whether the projects support local master plans, but on whether a city manages to be successful in bidding for an event. Mega-events thus become a wildcard, allowing cities to jump to the front of the queue for government support.

3. POLICY SUGGESTIONS

Cities, national governments, and governing bodies of events all have good reasons to reduce the negative fallout from mega-events. Cities, national governments, and citizens should want to avoid cost overruns, inefficient allocation of resources, and oversized infrastructure. Event-governing bodies, such as the IOC for the Olympic Games or FIFA for the Football World Cup, are concerned about their reputation, which constitutes their key capital. They also worry about losing potential hosts if cities start to perceive mega-events as expensive burdens. Among event-governing bodies, the IOC has been the most active in pursuing reforms, attempting to reduce the size of individual mega-events, integrate each one with urban development, and maximize the use of existing or -temporary infrastructure.

But why is it so difficult to offset the mega-event syndrome? First, the IOC and other event-governing bodies set the minimum size of infrastructure, but they cannot set the maximum size. Thus, even if the governing bodies set lower requirements, local organizers may not translate those requirements into smaller infrastructure investments and expenditures. Second, elites organize the bid for an event in most host countries and cities, precisely because they profit most from the events (see Logan & Molotch, 1987). Hosts continue to bid because of the unequal distribution of costs and benefits, despite negative outcomes for cities and regions as a whole. Third, the spectacular character of mega-events grips people and fires their imaginations, often sidelining rational deliberations about an event's benefits and costs, especially during the bidding phase. The fantasies attached to these events often turn out to be illusory the closer the event draws, but then it is too late for second thoughts.

At the policy level, as Table 2 illustrates, there are two avenues to address the mega-event syndrome. The first is *radical changes*: changes to the rules and legal preconditions of awarding, hosting, and funding mega-events. These -require a change in attitude among policymakers, city -administrations, event managers, and event-governing bodies. But these changes will often not occur without pressure, through media, social movements, referenda, or from corporate sponsors of the event. The second avenue to address the mega-event syndrome is through *incremental changes* to the current rules and policies. These policy changes are easier to implement, but they are also less effective.

Table 2. Policy changes for curbing the mega-event syndrome.

Radical Changes	Symptoms addressed	Incremental Changes	Symptoms addressed
Avoid tying mega-events to large-scale urban development	1, 2, 3, 5, 7	Seek public participation beginning in the bid stage	3, 4, 6
Bargain with event-governing bodies	2, 3, 4	Fix terms of hosting agreement at the time of bidding	3
Cap and earmark public expenditure	2, 3, 4, 7	Create a separate organization in charge of legacies	3, 4
Seek independent expert assessments	1, 2, 3	Decentralize the event	3
Reduce or cap size and requirements of events	1, 2, 3, 4, 5	Build temporary structures where afteruse is uncertain	3
		Engage in knowledge exchange	2, 3
		Do not bypass regular planning procedures	5, 6

Two main actors are responsible for implementing these changes—hosts and event-governing bodies—and in many cases both need to work together for a change to be most effective.

4. RADICAL CHANGES

4.1 Avoid Tying Mega-Events to Large-Scale Urban Development

Cities need to change the current trend of tying large-scale development projects to mega-events. They have to establish, before starting to bid, whether the events require extensive new construction or upgrading of existing infrastructure. If so, cities have three options: Bid for smaller events; build the required infrastructure before the bid, but only if it aligns with the master plan; or do not bid at all. Event-governing bodies could support this change by preferring bids with existing -infrastructure. The 1984 Summer Olympic Games in Los Angeles is one example where organizers relied to a large degree on existing infrastructure (see Burbank et al., 2002), thus reducing many symptoms of the mega-event -syndrome.

4.2 Bargain With Event-Governing Bodies

The IOC, FIFA, UEFA, and others are monopolies that can dictate their terms and make substantial risk-free income, as long as there is enough demand for their events. Host cities should attempt to gain concessions from event-governing bodies, including fewer requirements, full taxation of revenues, waiving government guarantees, or additional contributions to cover the cost of hosting.

But one of two things needs to happen to allow better bargaining with the governing bodies. The first is that interest in bidding for an event needs to drop substantially, which is what happened after the excesses of the 1976 Olympic Games in MontrÈal, when Los Angeles was the only bidder for the 1984 Olympics. As a result, Los Angeles was able to wring significant concessions from the IOC.

Second, those interested in bidding must band -together to bargain with event-governing bodies for –better conditions, thus contesting monopoly power. If this collective bargaining does not happen, interested bidders could negotiate individually. Oslo, a bidder for the 2022 Winter Games, was able to win concessions from the IOC to bear part of its costs (Butler, 2014). However, the success of bargaining depends on the willingness of the event-governing body to make concessions, the demand for the event, and the bargaining power of the host. Large markets such as the United States will exercise more power than smaller -markets such as South Korea.

4.3 Cap and Earmark Public Expenditure

The hosts should cap expenditures and earmark the funds to avoid having the public sector compensate for mega-event cost overruns and ventures that lose money. Capping expenditures reduces the risk of profiteering and overspending; earmarking prevents funds for urban development from being diverted to hosting the event itself. Host cities should involve the private sector in risk taking to ensure the commercial viability of facilities and to reduce the exposure of the public sector. Cities and governments should not give blanket guarantees to cover all costs.

In addion, national governments should not provide extra funding for urban development to support mega-events; this encourages bidding for mega-events just for the sake of extracting these funds. Making funding decisions this way raises the total cost of infrastructure delivery to society and perverts other ways of determining funding priorities, such as regional and national infrastructure planning processes. Of course, the degree to which a host can enforce a spending cap will depend on legal and -economic circumstances.

4.4 Seek Independent Expert Assessments

Independent expert advice is crucial for decision–makers in the cities, national governments, organizing committees, and governing bodies of the event. Such advice could take the form of reference class forecasting, a method that compares the predicted costs and benefits of a large number of mega-events—the so-called reference class—with the actual ones after the events have taken place. This approach determines how much predicted and actual costs and benefits diverged, and provides a better assessment of the bidding documents (Lovallo & -Kahneman, 2003).

4.5 Reduce or Cap Size and Requirements of Events

Reversing the constant growth of mega-events would reduce the size of the required venues and infrastructure and thus the risk of event takeover, the complexity of the management and thus the risk of cost overruns and benefit shortfalls, and the size of the overall building program and thus the necessity to introduce extralegal measures to complete it in time. The event-governing bodies could scale down the event by reducing the number of athletes and sports, the number of media, or the number of -visitors. Reducing the number of visitors appears to be the most viable option, while specific sports or events could be included on a rotational basis or made to share venues.

4.6 Incremental Changes

4.6.1 Seek Public Participation Beginning in the Bid Stage

Citizen involvement reduces the risk that mega-event priorities will take precedence over urban development priorities and ensures that citizens can have a say in mega-event planning. Public hearings and planning consultations with stakeholders, as Vancouver held for the 2010 Olympic Games, not only facilitate a better alignment of infrastructure with citizen needs but also build consensus and reduce potential opposition. Organizers or citizens can call a public referendum (in jurisdictions where legislation -provides for this) to let the population decide whether or not to host the event. Thus, the promoters of Boston's bid for the 2024 Summer Olympics decided to organize a referendum after public support had dropped to just 36% (Seelye,2015)

4.6.2 Fix Terms of Hosting Agreement at the Time of Bidding

Hosts should avoid signing any contracts that leave requirements deliberately vague or that postpone concrete specifications to a later stage. This vagueness makes it difficult for cities and countries to plan and budget in advance. Oslo, in its bid for the 2022 Winter Games, successfully insisted that the IOC would not retroactively introduce new requirements that would lead to higher costs for the city (Butler, 2014).

4.6.3 Create a Separate Organization in Charge of Legacies

A separate organization must ensure that what is left after the event—the so-called legacies—contributes to the long-term development of a city and region. This organization should be created at the bid stage and have a say in all matters of planning that reach beyond the event. This activity should have clear funding sources at the time of the bid. London has such an organization, but it was created after the bid was won; Vancouver's comparable organization only had responsibilities for the the social dimension of legacies.

4.6.4 Decentralize the Event

It makes more sense to spread demand spatially rather than build permanent facilities to accommodate peak demand. Events that take place in one city mean a few days or weeks of intense strain for public transport and accommodation. Instead of building many venues in the same place, venues should be distributed across a city or perhaps even to other cities. Los Angeles in 1984 and Vancouver in 2010 practiced such a -decentralized model and thus avoided building excessive transport capacity (see Liao & Pitts, 2006).

4.6.5 Build Temporary Structures Where Afteruse Is Uncertain

Building temporary facilities such as sports venues can both be cheaper than building permanent facilities and eliminate maintenance costs after the event for facilities that are hardly used or would otherwise be too large. Construction costs of temporary venues are between one-half and two-thirds of those of permanent venues (Long, 2013). London made extensive use of temporary venues for the 2012 Summer Games, where organizers built 11 of the 34 competition venues as temporary structures from scratch (May & Cardwell, 2012). There is, however, a drawback to temporary structures: They increase event-specific expenditure that is unproductive for urban -development, so hosts must weigh the costs and benefits in each case.

4.6.6 Engage in Knowledge Exchange

Better knowledge exchange among past and future hosts can alleviate uncertainty in the demand for infrastructure and services during the event. It can also avoid reinventing the wheel where efficient solutions are available elsewhere, thus reducing budget and time overruns. While the IOC and FIFA have started knowledge transfer programs, these are focused on the organizing committees, which are in charge of the operation of the event. To be effective, - however, such programs also have to encompass the -organizations responsible for delivering the infrastructure and other services for the event.

4.6.7 Do Not Bypass Regular Planning Procedures

Regular planning procedures should remain in force for mega-events, even though their fixed deadlines increase time pressures. These procedures exist to ensure equal consideration of interests, rational decision making, and fair tendering and bidding. To override regular procedures increases the risk of nontransparent decisions that favor certain stakeholders over others.

5. CONCLUSION

In this study, I make the case that there is a discernable mega-event syndrome: a set of seven major symptoms that afflict mega-event planning. Together, these symptoms too often turn mega-events into obstacles rather than boons to urban development. The mega-event syndrome results in oversized or obsolete infrastructure for an inflated price that the public is forced to pay and in an uneven and inefficient allocation of resources. These symptoms repeat themselves, to a greater or lesser degree, in mega-events around the globe. I suggest some incremental changes that can improve the outcomes of mega-events for local hosts, but the more marginal changes can only go so far in -reducing the negative impacts on hosts.

Radical changes to the rules of the games—in how mega-events are planned, awarded, and governed—are needed. These more radical changes may require outside pressure from a critical public, social movements, NGOs, the media, or corporate sponsors. In recent years, we have witnessed an escalating spiral in which ever-greater requirements and costs for mega-events have compelled cities to promise ever-greater projects. The most important radical change cities could adopt is to stop bundling mega-event planning with large-scale urban development, which comes with a higher risk of cost overruns, substandard construction quality, oversized infrastructure, and a lack of -democratic participation. Tying development projects to mega-events makes the already complex planning for mega-events even more complex because it increases the number of interdependent elements.

There has been a marked decline in communities willing to host mega-events driven by the record of recent mega-events that provided limited benefits at exorbitant costs, such as the 2014 World Cup in Brazil and the 2014 -Winter Games in Sochi. The 2022 Winter Olympics have turned into the "Games that no one seems to want" (-Wilson, 2014), after Munich, Oslo, Stockholm, St. Moritz/Davos, Krakow, and Lviv refrained from -submitting bids or withdrew them. The remaining two contenders in the race—Kazakhstan and China—could portend the future of mega-events as lavish image projects in resource-rich, autocratic states with little public -accountability and limited freedom of speech. The bids for the 2024 Olympic Summer Games and the 2026 Football World Cup are now under way, and the United States stands a high chance of winning both if it decides to bid. Debates and actions to counteract the mega-event syndrome must start now, early in the bid phase, when the basic parameters can still be changed.

ACKNOWLEDGMENTS

Thanks are due to audiences in Brussels, Kazan, Rio de Janeiro, and Zurich for critical feedback on earlier versions, and to Christopher

Gaffney and Daniel Wolfe for commenting on this manuscript. The three anonymous reviewers helped shape this article into what it is now: Thank you for your time and effort. Part of this article was completed while on a visiting fellowship with the Brussels Centre for Urban Studies at the Vrije Universiteit Brussel: Thank you for your hospitality.

NOTES

1 Event-governing bodies argue that the use of public funds for non-sports-related infrastructure, such as for transport or communications, should be seen as investment and thus not counted as costs of a mega-event. This line of argument, however, is only partly correct. The problem is that, as seen before, mega-events may change what infrastructure is built—what is required for the event and not what has the greatest public utility—and its size. In addition, mega-events inflate the price tag for infrastructure through the mechanisms discussed in Symptom 2. As such, the public still risks paying for overpriced but underused infrastructure.

2 The displacement of 700,000 residents in the run-up to the 1988 Olympic Games in Seoul and more than a million residents for the 2008 Beijing Olympic Games, however, must also be seen as situations where events were part and parcel of a larger redevelopment strategy and thus not the only driver and cause of displacement (Davis, 2011).

3 However, as mentioned earlier, some elites profit from this inefficient allocation of resources, so there is an incentive for them to overbuild. Some policy changes discussed later in this study can reduce this incentive.

4 In 2003, the IOC published a report condemning the growing size of the Olympic Games and issued more than 100 mostly operational

recommendations to counteract this trend (Olympic Games Study Commission, 2003). The IOC also asks for bid books to detail how Olympic planning fits into the long-term development plans of the host, and IOC members are asked to take this into account when voting for a host. Beginning in 2001, it standardized the organization of the event and introduced a knowledge transfer system to make organizational processes more efficient. Despite past reforms, however, the IOC has been unable to reverse, stop, or even noticeably slow the growth of the Olympics (Chappelet, 2014). The Olympic Agenda 2020, passed in December 2014, steps up efforts to enhance transparency, reducing the cost and size and maximizing the use of existing or temporary infrastructure. It yet has to show whether it can be more effective than its predecessors.

5 As an adverse effect, this implication may encourage investments into infrastructure in anticipation of increasing chances to win an event, but these investments would then occur without the fixed deadline of the event, the state of legal exception, and a number of other symptoms, and it would be harder to justify event-specific expenditure vis-à-vis citizens if an event has not been awarded.

6 The IOC has recognized the need for keeping the terms of the hosting agreement stable and for decentralization in its Agenda 2020, adopted in December 2014 (Recommendation 1).

REFERENCES

1 Andranovich, G., Burbank, M. J., & Heying, C. H. (2001). Olympic cities: Lessons learned from mega-event politics. Journal of Urban Affairs, 23(2), 113–131.

2 Baade, R. A., & Matheson, V. (2004). The quest for the Cup: Assessing the economic impact of the World Cup. Regional Studies, 38(4), 343–354.

3 Baumann, R., & Matheson, V. (2013). Infrastructure investments and mega-sport events: Comparing the experience of developing and industrialized countries (Faculty Research Series Paper No. 13-05). Worcester, MA: College of the Holy Cross, Department of Economics.

4 Belisário, A. (2014, July 1). Quatro Irmãs: assim atua o capitalismo brasileiro [The Four Sisters: This is how Brazilian capitalism works]. CartaCapital. Retrieved from http://outraspalavras.net/brasil/as-quatro-irmas-uma-historia-do-capitalismo-brasileiro/

5 Boykoff, J. (2011). Space matters: The 2010 Winter Olympics and its discontents. Human Geography, 4(2), 48–60.http://www.hugeog.com/

6 Boykoff, J. (2014). Celebration capitalism and the Olympic games. London, UK: Routledge.

7 Burbank, M. J., Andranovich, G., & Heying, C. H. (2002). Mega-events, urban development and public policy. Review of Policy Research, 19(3), 179–202.

8 Butler, N. (2014, September 23). Oslo 2022 claim IOC changes to Host City contract "well-received." Retrieved fromhttp:// www.insidethegames.biz/olympics/winter-olympics/2022/1022771-exclusive-oslo-2022-claim-ioc-changes-to-host-city-contract-well-received

9 Centre on Housing Rights and Evictions. (2007). Fair play for housing rights. Geneva, Switzerland: Author.

10 Chalip, L. (2006). Towards social leverage of sport events. Journal of Sport & Tourism, 11(2), 109–127.

11 Chappelet, J.-L. (2014). Managing the size of the Olympic Games. Sport in Society, 17(5), 581–592.

12 Coaffee, J. (2014). The uneven geographies of the Olympic carceral: From exceptionalism to normalisation. The Geographical Journal. Advance online publication.

13 Cottle, E. (Ed.). (2011). South Africa's World Cup: A legacy for whom? Scottsville, South Africa: University of KwaZulu-Natal Press.

14 Crompton, J. L. (1995). Economic impact analysis of sports facilities and events: Eleven sources of misapplication. Journal of Sport Management, 9(1), 14–35. Retrieved fromhttp:// www.humankinetics.com/acucustom/sitename/Documents/DocumentItem/7669.pdf

15 Davies, G. A. (2008, April 25). Mayor tricked government into 2012 Olympics bid. The Telegraph. Retrieved fromhttp:// www.telegraph.co.uk/sport/olympics/2298374/Mayor-tricked-Govt.-into-2012-Olympics-bid.html

16 Davis, L. K. (2011). International events and mass evictions: A longer view. International Journal of Urban and Regional Research, 35(3), 582–599.

17 de Paula, M. (2014). The 2014 World Cup in Brazil: Its legacy and challenges. São Paulo, Brazil: Heinrich-Böll-Stiftung.

18 Eisinger, P. (2000). The politics of bread and circuses: Building the city for the visitor class. Urban Affairs Review, 35(3), 316–333.

19 Essex, S., & Chalkley, B. (1998). Olympic Games: Catalyst of urban change. Leisure Studies, 17(3), 187–206.

20 Flyvbjerg, B., Holm, M. S., & Buhl, S. (2002). Underestimating costs in public works projects: Error or lie? Journal of the American Planning Association, 68(3), 279–295.

21 Flyvbjerg, B., & Stewart, A. (2012). Olympic proportions: Cost and cost overrun at the Olympics 1960–2012 (Working Paper). Oxford, UK: University of Oxford Said Business School. Retrieved from http://papers.ssrn.com/sol3/papers. cfm?abstract_id= 2238053

22 French, S. P., & Disher, M. E. (1997). Atlanta and the Olympics: A one-year retrospective. Journal of the American Planning Association, 63(3), 379–392.

23 Fund for the Fight Against Corruption. (2014). Entsiklopediya trat Sochi-2014. Moscow, Russia: Author.

24 Gaffney, C. (2010). Mega-events and socio-spatial dynamics in Rio de Janeiro, 1919–2016. Journal of Latin American Geography, 9(1), 7–28.

25 Gaffney, C. (2014). A World Cup for whom? The impact of the 2014 World Cup on Brazilian football stadiums and cultures. In P. Fontes & B. Buarque (Eds.), The country of football: Politics, popular culture, and the beautiful game in Brazil (pp. 189–208). London, UK: Hurst.

26 German Development Agency. (2013). Groflveranstaltungen als Motor fr nachhaltige Entwicklung [Mega-events as engines of sustainable development]. Frankfurt, Germany: GIZ.

27 Gold, J. R., & Gold, M. M. (2008). Olympic cities: Regeneration, city rebranding and changing urban agendas. Geography Compass, 2(1), 300–318.

28 Gold, M. M. (2007). Athens 2004. In J. R. Gold & M. M. Gold (Eds.), Olympic cities: City agendas, planning and the world's games, 1896–2012 (pp. 265–285). Abingdon, UK: Routledge.

29 Gratton, C., Shibli, S., & Coleman, R. (2005). Sport and economic regeneration in cities. Urban Studies, 42(5–6), 985–999.

30 Grix, J. (2012). "Image" leveraging and sports mega-events: Germany and the 2006 FIFA World Cup. Journal of Sport & Tourism, 17(4), 289–312.

31 Hall, C. M. (2006). Urban entrepreneurship, corporate interests and sports mega-events: The thin policies of competitiveness within the hard outcomes of neoliberalism. In J. Horne & W. Manzenreiter (Eds.), Sports mega-events: Social scientific analyses of a global phenomenon (pp. 59–70). Oxford, UK: Blackwell.

32 Hayes, G., & Horne, J. (2011). Sustainable development, shock and awe? London 2012 and civil society. Sociology, 45(5), 749–764.

33 High Level Committee for Commonwealth Games 2010. (2011). Organization and conduct of Commonwealth Games 2010. New Delhi, India: Author.

34 Hiller, H. (2000). Mega-events, urban boosterism and growth - strategies: An analysis of the legitimations of the Cape Town 2004 Olympic bid. International Journal of Urban and Regional Research, 24(2), 439–458.

35 Hiller, H. (2006). Post-event outcomes and the post-modern turn: The Olympics and urban transformations. European Sport Management Quarterly, 6(4), 317–332.

36 Holt, R., & Ruta, D. (2015). Routledge handbook of sport and legacy: Meeting the challenge of major sports events. London, UK: Routledge.

37 Horne, J. (2007). The four "knowns" of sports mega-events. Leisure Studies, 26(1), 81–96.

38 Jago, L., Dwyer, L., Lipman, G., van Lill, D., & Vorster, S. (2012). Optimising the potential of mega-events: An overview.

International Journal of Event and Festival Management, 1(3), 220–237.

39 Jennings, W. (2012). Olympic risks. Basingstoke, UK: Palgrave Macmillan.

40 Kantar. (2014). IOC research: Sochi 2014. Chambourcy, France: Kantar Media. Retrieved fromhttp://www.olympic.org/Documents/Games_Sochi_2014/Research-directive-Kantar.pdf

41 Kassens-Noor, E. (2012). Planning Olympic legacies: Transport dreams and urban realities. London, UK: Routledge.

42 Lenskyj, H. (2008). Olympic industry resistance: Challenging Olympic power and propaganda. Albany: SUNY Press.

43 Levenson, M. (2015, January 21). Boston city employees told not to badmouth Olympics. Boston Globe. Retrieved fromhttps://www.bostonglobe.com/metro/2015/01/21/boston-city-employees-told-not-badmouth-olympics/ J7JFBrq2YGkWGzqP42PQZO/story.html

44 Liao, H., & Pitts, A. (2006). A brief historical review of Olympic urbanization. The International Journal of the History of Sport,23(7), 1232–1252.

45 Logan, J. R., & Molotch, H. L. (1987). Urban fortunes: The political economy of place. Berkeley: University of California Press.

46 Long, J. G. (2012). Public/private partnerships for major league sports facilities. London, UK: Routledge.

47 Long, J. G. (2013). Rethinking Olympic infrastructure. Retrieved from http://lsecities.net/media/objects/articles/rethinking-olympic- infrastructure/en-gb/

48 Louw, A. M. (2012). Ambush marketing & the mega-event monopoly: How laws are abused to protect commercial rights. The Hague, The Netherlands: Asser Press.

49 Lovallo, D., & Kahneman, D. (2003, July). Delusions of success: How optimism undermines executives' decisions. Harvard Business Review, 56–63. Retrieved from http://hbr.org

50 Madden, J. R. (2006). Economic and fiscal impacts of mega sporting events: A general equilibrium assessment. Public Finance and Mangement, 6(3), 346–394. http://www.spaef.com/pfm.php

51 Matheson, V. (2008). Mega-events: The effect of the world's biggest sporting events on local, regional, and national economies. In D. Howard & B. Humphrey (Eds.), The business of sports (Vol. 1, pp. 81–99). -Westport, CT: Praeger.

52 May, P., & Cardwell, S. (2012). Temporary Olympic structures. Ingenia, (51), 2–7. http://www.ingenia.org.uk

53 McMillan, J. H., & Schumacher, S. (1997). Research in education: A conceptual introduction. New York, NY: Longman.

54 Mills, B. M., & Rosentraub, M. S. (2013). Hosting mega-events: A guide to the evaluation of development effects in integrated metropolitan regions. Tourism Management, 34(1), 238–246.

55 Nuzman, C. (2011, January). Os Projetos Dos Jogos Rio 2016 [The projects of the Olympics in Rio 2016]. Paper presented at the press conference at the AssociaÁ„o dos Correspondentes da Imprensa -Estrangeira, Rio de Janeiro.

56 Olds, K. (1998). Urban mega-events, evictions and housing rights: TheCanadian case. Current Issues in Tourism, 1(1), 2–46.

57 Olympic Games Study Commission. (2003). Final report to the 115th IOC session. Lausanne, Switzerland: IOC.

58 Pletz, J. (2009, August 24). Daley's Olympian stretch. Crain's - Chicago Business. Retrieved fromhttp://www.chicagobusiness.com/article/20090822/ISSUE01/100032302/daleys-olympian-stretch

59 Porter, L., Jaconelli, M., Cheyne, J., Eby, D., & Wagenaar, H. (2009). Planning displacement: The real legacy of major sporting events. Planning Theory & Practice, 10(3), 395–418.

60 Preuss, H. (2004). The economics of staging the Olympics: A comparison of the Games, 1972–008. Cheltenham, UK: Edward Elgar.

61 PricewaterhouseCoopers. (2011). The Games effect. Toronto, ON, Canada: Author.

62 Priemus, H., Flyvbjerg, B., & van Wee, B. (Eds.). (2008). Decision-making on mega-projects: Cost-benefit analysis, planning, and innovation. Cheltenham, UK: Edward Elgar.

63 Raco, M. (2014). Delivering flagship projects in an era of regulatory capitalism: State-led privatization and the London Olympics 2012. International Journal of Urban and Regional Research 38(1), 176–197.

64 Reuters. (2013, April 24). Soccer: Less democracy makes for an easier World Cup—Valcke. Reuters. Zurich. Retrieved fromhttp://www.reuters.com/article/2013/04/24/us-soccer-fifa-idUSBRE93N18F20130424

65 Rosentraub, M. S. (1999). Major league losers: The real cost of sports and who's paying for it (rev. ed.). New York, NY: Basic Books.

66 Rutheiser, C. (1997). Making place in the nonplace urban realm: Notes on the revitalization of downtown Atlanta. Urban Anthropology and Studies of Cultural Systems and World

Economic Development, 26(1), 9–42. Retrieved from http://the-institute-ny.com/WEB%20PAGES/JOURNAL/uasjournal.htm

67 Sánchez, F., & Broudehoux, A.-M. (2013). Mega-events and urban regeneration in Rio de Janeiro: Planning in a state of emergency. International Journal of Urban Sustainable Development, 5(2), 132–153. doi:10.1080/19463138.2013.839450

68 Schellnhuber, H.-J., Block, A., Cassel-Gintz, M., Kropp, J., Lammel, G, Lass, W., ÖReusswig, F. (1997). Syndromes of global change. GAIA, 6(1), 19–34. Retrieved from http://www.pik-potsdam.de/~kropp/paper/journals/gaia6_1_1997.pdf

69 Scherer, J. (2011). Olympic villages and large-scale urban development: Crises of capitalism, deficits of democracy?Sociology, 45(5), 782–797.

70 Seelye, K. Q. (2015, March 24). In reversal, Boston 2024 wants vote on Olympic bid. The New York Times. Retrieved fromhttp://www.nytimes.com/2015/03/25/sports/olympics/in-reversal-boston-2024-wants-a-vote-on-the-olympics.html

71 Shaw, C. A. (2012). The economics and marketing of the Olympic Games from bid phase to aftermath. In H. J. Lenskyj & S. Wagg (Eds.), The Palgrave handbook of Olympic studies (pp. 248–260). New York, NY: Palgrave.

72 Shin, H. B., & Li, B. (2013). Whose games? The costs of being "Olympic citizens" in Beijing. Environment and Urbanization,25 (2), 559–576. doi:10.1177/0956247813501139 ,

73 Teigland, J. (1999). Mega-events and impacts on tourism: The predictions and realities of the Lillehammer Olympics. Impact Assessment and Project Appraisal, 17(4), 305–317.

74 Tversky, A., & Kahneman, D. (1974). Judgment under uncertainty: Heuristics and biases. Science, 185(4157), 1124–1131.

75 Watt, P. (2013). "It's not for us': Regeneration, the 2012 Olympics and the gentrification of East London. City, 17(1), 99–118.

76 Wilson, S. (2014, June 30). Ukraine's Lviv withdraws bid for 2022 Winter Games. Associated Press. Retrieved fromhttp://sports.yahoo.com/news/ukraines-lviv-withdraws-bid-2022-071137232--spt.html

77 Zimbalist, A. (2015). Circus Maximus: The economic gamble behind hosting the Olympics and the World Cup. Washington, DC: Brookings Institution Press.

CHAPTER 10

Service quality assessment of transportation and government services: a study of the Hong Kong tourism industry

Eddy K. Tukamushaba[1] Babu P. George[2]

[1]School of Hotel and Tourism Management, The Hong Kong Polytechnic University, Office Room: HOTEL ICON RM 842, 17 Science Museum Road, Tsim Sha Tsui East, Hung Hom, Kowloon, Hong Kong

[2]Department of Business Administration, Alaska Pacific University, 4790 University Drive #A, Anchorage, AK-99508, USA

ABSTRACT

This paper presents the results of a study examining service quality assessment of selected sectors of the Hong Kong tourism industry. The functioning of three related sectors of air, rail, and road transport, as well as tourism related government departments, is considered. Identification of the visitors and their perceptions of quality of service offered by these sectors are examined. In order to establish the importance of each of the service quality attributes in determining each sector's performance, importance-performance analysis was conducted on the airlines, public transport, and government agencies such as police, immigration, customs, and leisure and cultural services. The implications of the results obtained are discussed.

KEYWORDS

service quality, satisfaction, loyalty, tourism, government services, transportation, Hong Kong

1 INTRODUCTION

Tourist satisfaction is increasingly becoming an important area of concern for many sectors that depend on inbound tourism. This is not an exception to the Hong Kong tourism industry as Hong Kong relies heavily on the contribution from the tourism industry to the tune of HK$228.8bn or 11.9% of GDP in 2011 (WTTC, 2011).

However, except in attraction and hotel sectors, not many empirical studies have been carried out to establish how specific tourism subsectors in Hong Kong are performing in various service quality attributes in recent years (Song et al., 2011). Moreover, the identification of visitors' perceptions of the quality of service offered by the Hong Kong tourism experience in related sectors especially transport has not been fully explored (Jin et al., 2008; Song et al., 2008). This information is vital when performance improvement is required to enable the provision of quality service to match up with the tourists' expectations. Though some research has been conducted in the areas of transportation in general (Chew, 1987; Khadaroo and Seetanah, 2008; Prideaux, 2000), we note that airline and road-rail public transport contributions to tourism development in Hong Kong are two sectors that have not got enough academic attention from tourism focused researchers. As the proverbial saying goes, a chain cannot be stronger than its weakest link: tourism is the result of a complex chain of activities, and overall service performance cannot be assessed by overly concentrating on only one or two top industry segments (George, 2005).

Johnston (1995) notes that measuring and aggregating performance in all areas that constitute a service is the key to understand customer satisfaction and loyalty. According to Ajzen and Driver (1992), customers' willingness to re-purchase, positive word of mouth, and ability to recommend the services stem from overall satisfaction. Consumers think holistically and increase in overall service performance has also been found to lead to an increase in perceptions of service quality (Gronroos, 1993; Johnston and Lyth, 1998). Enhancing service quality through strategic planning and the appropriate allocation of resources to various sectors of the tourism industry is important for improving destination competitiveness and tourism growth (Crouch and Ritchie, 1999; Gomezelj and Mihalic, 2008; Poon, 1993; Ritchie and Crouch, 2003; Yoon et al., 2001).

In this background, the present researchers investigate some of the service quality related issues in the specific context of transportation (road, rail, air) and government services. Segments of visitors and their perceptions of service quality offered by road, rail, and air transport sectors are examined. In order to establish the importance of service quality attributes in determining each sector's performance, importance-performance analysis (IPA) was conducted on the transport sectors and government agencies such as police, immigration, customs, and leisure and cultural services.

The rest of the paper is organised as follows. The current Section 1 provides the background and rationale for this study while Section 2 provides a literature review and conceptual framework. Section 3 addresses the methodology used and Section 4 presents findings and discussion. The last section provides conclusions and recommendations.

2 CONCEPTUAL OVERVIEW

In the review of literature presented below, we will highlight and summarise the established relationships among some of the key constructs in the consumer behaviour literature. A schema for the review is given below in Figure 1.

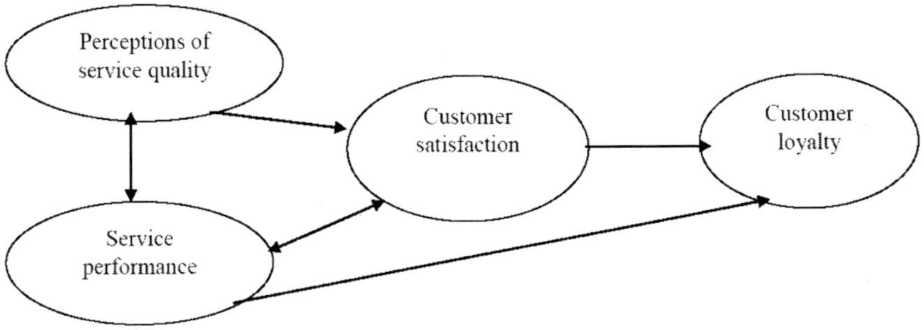

Figure 1 Conceptual framework used for literature review

2.1 Expectation-satisfaction-loyalty-service performance linkage

It is evident in various studies that satisfaction determines the post-purchase decision (Boulding et al., 1993; Dabholkar and Thorpe,

1994; Fornell, 1992; Oliver and Swan, 1989; Keaveney, 1995). While positive satisfaction will definitely lead to repurchase intentions (Gotlieb et al., 1994), the opposite also holds true: dissatisfied visitors are unlikely to visit a destination again (Dube et al., 1994). Hence, 'satisfaction' is an effective indicator to predict and evaluate the intention of a customer to repurchase (Choi and Chu, 2001; Petrick, 2002). In addition, one way to obtain higher profit is to retain regular guests because the cost of keeping a loyal customer is lower than that of attracting a new one (Kim et al., 2009; Lee et al., 2000; Park et al., 2004). In the study of Gupta et al. (2004), it is found that a 1% increase in customer retention rate results in a 5% increase in profits.

It is commonly known that consumer expectations affect satisfaction. The higher the expectation, the more likely the disappointment is. Expectations are formed by personal experience, word-of-mouth, personal needs, and marketing of the product/service (Parasuraman et al., 1985). Several theories have been adopted for measuring customer satisfaction. The most widely preferred one is the *expectancy disconfirmation paradigm*(EDP) introduced by Oliver (1980), in which the actual performance is measured against expectation. If the expectation of the consumers is met, confirmation occurs. However, if the expectation is unmet, it could result in a positive disconfirmation if the performance is better than expectation or a negative disconfirmation if the performance is below expectation.

Another model built on EDP is *SERVQUAL* introduced by Parasuraman et al. (1988). The concept of the model is to examine customer satisfaction by measuring the gap between perceived expectation and service performance based on five service dimensions: *tangibles, reliability, responsiveness, assurance, empathy.* Knowing this gap is of critical importance for service improvement so as to retain customers. Various studies (Howat et al., 1996; Hui et al., 2007; Armstrong et al., 1997; Hsieh et al., 2008; Atilgan et al.,

2003) have applied SERVQUAL to assess customer satisfaction in the tourism industry. Other than the research of Thompson and Schofield (2007) which looked at the public transport sector and the study by Choi and Chu (2001) and Saleh and Ryan (1992) which investigated the hotel industry, the work of Kim and Lee (2011) found that to low cost carriers the two dimensions of 'tangibles' and 'responsiveness' are crucial for enhancing customer satisfaction and behavioural intentions. Moreover, Gilbert and Wong (2003) noticed that the dimension of 'assurance' is of paramount importance to the success of airline services.

However, the model of SERVQUAL is not without limitations. One of these is that customers tend to expect high level of service (Brown et al., 1993; Qu and Sit, 2007), thus it is difficult to match up with this expectation even though the performance of the service providers is good. Besides, the service attributes to evaluate the service performance may not cover all the important characteristics of a particular service (Akama and Kieti, 2003). Nonetheless, SERVQUAL is still considered an effective model to examine the potential issues related to the service performance from the view of a service provider and management.

Therefore, in order to explore and understand the close relationship between loyalty, customer satisfaction, expectation and service performance, the model of SERVQUAL is applied in this study as a basis to examine the service performance of the tourism sectors of airline, government service and public transport for which statistics were obtained.

2.2 Attributes of customer satisfaction

Customer satisfaction can be measured through the attributes experienced by the customers (Churchill and Surprenant, 1982), and customer satisfaction is a function of both expectations of the attributes and the perceived performance of such attributes of the products or services (Martilla and James, 1977). Martilla and James (1977) introduced a technique called 'IPA' which measures customer satisfaction based on the actual performance of various attributes and the customers' perceptions of the importance of such attributes. With IPA, the importance and performance of the same attributes can then simultaneously be compared. The concept of IPA is to plot the results in a two-dimensional matrix with the means of importance of various attributes along the x-axis, and the means of performance of such attributes along the y-axis. The matrix is categorised into four quadrants, namely 'Concentrate Here', 'Keep Up the Good Work', 'Low Priority', and 'Possible Overkill'. The results of IPA provide an attractively succinct visual snapshot of how the company performs in each attribute according to the customers' importance concerns (Haemoon, 2000), which helps a firm to identify those attributes that are the drivers of customer satisfaction (Deng, 2007). Hansen and Bush (1999) suggested that IPA can assist the practitioners to prioritise the improvement areas and to direct marketing strategies. Furthermore, the results provide guidelines for how organisations can deploy future resources in a better way in order to achieve a higher level of customer satisfaction (Deng, 2007; Haemoon, 2000).

Visitors' experiences and their satisfaction at destinations are often measured by examining various attributes and importance levels of such attributes of the tourism products and services provided by the destinations (Thompson and Schofield, 2007). Various researchers have adopted IPA as a research tool to analyse the tourists' satisfaction with the services and products provided by a destination. For instance, in the studies of Wilkins (2010), Chu and Choi (2000) and Qu and Sit (2007), the sector they examined with IPA was the hotel industry. Other examples include the study of Lacher and Harrill (2010) in which the authors examined the non-traditional *3S* activities of a *3S* destination; the researches of Jin et al.

(2008) and Song et al. (2008) in which they examined the service provided by airline, hotel and restaurant, and the works of Enright and Newton (2004) and Lee and Lee (2009) in which they investigated the general factors at large.

2.3 Importance of attribute and attribute satisfaction

Customer satisfaction is a function of both expectations of the attributes and the perceived performance of such attributes of the products or services (Martilla and James, 1977). Martilla and James (1977) introduced a technique called 'IPA', which measures the customer satisfaction based on the actual performance of various attributes and the customers' perceptions of the importance of such attributes. With IPA, the importance and performance of the same attributes can then simultaneously be compared. The concept of IPA is to plot the results in a two-dimensional matrix with the means of importance of various attributes along the x-axis, and the means of performance of such attributes along the y-axis. The matrix is categorised into four quadrants, namely 'concentrate here', 'keep up the good work', 'low priority', and 'possible overkill'. The results of IPA provide an attractive snapshot of how the company performs in each attribute according to the customers' importance concerns (Haemoon, 2000), in which it helps a firm to identify which attributes are the drivers of customer satisfaction (Deng, 2007). Hansen and Bush (1999) suggested that IPA can assist the practitioners to prioritise the improvement areas and to direct marketing strategies. Furthermore, the results provide guidelines of how organisations can deploy future resources in a better way in order to achieve a higher level of customer satisfaction (Deng, 2007; Haemoon, 2000).

Visitors' experiences and their satisfaction at destinations are often measured by examining various attributes and importance level of such attributes of the tourism products and services provided by the destinations (Thompson and Schofield, 2007). Various researchers have adopted IPA as a research tool to analyse the tourist satisfactions with the services and products provided by a destination. For instance, in the studies of Wilkins (2010), Chu and Choi (2000) and Qu and Sit (2007), the sector they examined with IPA was the hotel industry. Other examples include the study of Lacher and Harrill (2010) in which they examined the non-traditional *3S* activities of a *3S* destination, the researches of Jin et al. (2008) and Song et al. (2008) in which they examined the service provided by airline, hotel and restaurant, and the works of Enright and Newton (2004) and Lee and Lee (2009) in which they investigated the general factors in large. However, so far there have not been any researches using IPA to investigate visitor satisfaction of a destination regarding the services provided by airline, transportation, and government. Hence, one of the objectives of this study is to fill in this gap.

However, so far there has not been any research conducted using IPA to investigate visitor satisfaction with a destination regarding the services provided by airline, transportation, and government in Hong Kong. Hence, one of the objectives of this study is to fill in this gap.

3 METHODOLOGY

The visitor survey covering the quantitative aspects of the visitor satisfaction level towards the tourism industry and the relevant sectors in Hong Kong was conducted during the period 3rd December 2004 to 6th January 2005. A questionnaire approach was used to collect the information on visitors' expectations and their satisfaction with the service provided by the tourism-related sectors in Hong Kong.

3.1 Research design

The concept of service quality is conceptualised in the service quality literature on perceived quality, which is defined as the customers' judgement about a service provider's overall excellence. Perceived service quality can be measured by a comparison of expectations and the perceptions of the performance of different service attributes. In selecting the service quality dimensions/attributes to assess the overall service quality, it is important to note that these attributes should be regarded as important to visitors and contribute significantly to their assessment on service quality. A good knowledge of these attributes and the ability to measure them will help produce useful information for developing effective ways of improving service quality.

The service quality attributes used in this study were based on but not limited to Parasuraman et al. (1990), which identified five dimensions of service quality that include *tangibles* (physical facilities, equipment and appearance of personnel); *reliability* (ability to perform the promised service with accuracy and professionalism); *responsiveness* (willingness to help customers and to provide prompt service); *assurance*(understanding customers' needs and being courteous and able to transmit confidence to customers); and*empathy* (caring individual attention).

Based on these service dimensions, a questionnaire that covers all the relevant aspects was developed for all sectors under this study. The initial questionnaire for each sector contained around twenty attributes. However, due to the time constraint on the response to the number of questions in each questionnaire, the number of service attributes was reduced to around ten for each sector. The main reason for capping the number of questions is that empirical research suggests the ideal time length for completing a questionnaire is between 12 to 15 minutes. A short questionnaire does not allow sufficient data to be generated for in-depth analysis while a long questionnaire tends to result in a loss of concentration by respondents. Overall, an effort was made to cover, as extensively as possible, all the five broad dimensions of service quality as appropriate.

3.2 Sampling method

A multi-stage sampling approach was used in the data collection. In the first stage, the researchers chose the locations (strata) for interviews. In line with Tourism Commission's recommendation, the following locations were selected to conduct the survey. These locations are

1 Hong Kong International Airport

2 Macau Ferry Terminal

3 China Ferry Terminal

4 KCR Hung Hom Station

5 the Avenue of Stars.

The second stage is to use the convenient sampling approach to interview visitors. Different sample sizes with valid responses were obtained for each of the sectors because of location and traffic patterns. For final analysis a total number of 354 respondents for airlines, 362 for railway, and 304 government services were used.

3.3 Measuring service quality

Tourists usually have their initial expectations of the type and quality of services to be offered by the service provider. The extent to which tourists' expectations are met determines the level of tourists' satisfaction. Therefore, the questionnaire design of this study needed to reflect both the visitors' expectations as well as their satisfaction level of the service attributes in order to assess service quality. In the questionnaire survey, the visitors were asked to rate the importance (expectation) of the service attributes identified for each of the sectors using a five-point scale with 1 being not important all, 2 not important, 3 neutral, 4 important and 5 very important. At the same time the visitor was asked to rate their satisfaction level with these attributes using another five-point scale with 1 being very dissatisfied, 2 dissatisfied, 3 neither dissatisfied nor satisfied, 4 satisfied and 5 very satisfied.

Data analysis was conducted using SPSS version 18. Exploratory factor analysis using principal axis factoring with oblique (Direct-Oblimin) rotation was carried out to ensure that factors obtained were those reflected by the items used in this study (the Appendix). This process helped in deleting questions that had low factor loading (>.50). Reliability tests were conducted to establish Cronbach alpha coefficients for the scales in the research instrument. The government had four subscales of police consisting of six items (α = .774), immigration with five items (α = .781), customs with seven items (α = .861), and leisure and cultural services – 18 items (α = .919). That for airlines had ten items (α = .814), while transport had three subscales of railways with 12 items (α = .865), franchised buses with ten items (α = .878), and taxis eight items (α = .836). These findings indicate that all the scales were very good because they all had Cronbach's alpha coefficient greater than .7 and therefore were reliable (Nunnaly, 1978).

Descriptive analysis was conducted to obtain the measures of central tendency like means and standard deviations while frequency distributions were used to obtain the general demographic characteristics of the respondents, and cross tabulations were used to identify categorical variables that had a relationship with the visitors' intention to use services offered in the tourism sub sectors considered in the study, which indirectly provided an indication of the perceptions of the quality of service in the three sectors of the tourism industry.

In order to ascertain how the tourism subsectors were performing in providing quality service to match up with the tourists' expectations, the IPA was used while cluster and discriminant analyses were used to provide useful quantitative information in terms of areas that were contributing towards visitor satisfaction and thereby providing a basis for recommending service quality improvement to the relevant sectors.

4 ANALYSIS AND FINDINGS

This section provides the results from the different analyses that were conducted to identify the visitors and their perceptions of quality of service offered by the tourism and related industries, to ascertain how the industries were performing in providing quality service to match up with the tourists' expectation, with quantitative information provided for each of the sectors.

4.1 General findings

As mentioned, a range of locations (multi-stage sampling) commonly used by the Hong Kong Tourism Board for such surveys was used in the data collection. For the airline sector, Hong Kong International Airport accounted for 88.7%, China Ferry terminal 3.1%, KCRC Hung Hum 1.7% and Avenue of Stars 6.5%, with a total sample of 354 respondents. For the government sector a total sample of 304 valid responses was obtained with Hong Kong International Airport accounting for 49.3%, China Ferry terminal 4.3%, Macau ferry terminal 12.2%, KCRC Hung Hum 4.6% and Avenue of Stars 29.6%. Lastly for the Transport sector, a total of 362 respondents were surveyed and Hong Kong International Airport accounted for 41.1%, China Ferry terminal 13.8%, Macau ferry terminal 9.9%, KCRC Hung Hum 9.1% and Avenue of Stars 19.1%. All of the data was collected during the period from 3rd December 2004 to 6th January 2005. Table 1 shows that the distribution of respondents by region for the three sectors.

Table 1 The regional distribution of respondents was as follows

Region	*Airlines (%)*	*Government (%)*	*Transport (%)*
Mainland China	34.5	34.2	35.4
Taiwan	14.1	13.2	13.0
South and Southeast Asia	11.6	11.2	10.5
North Asia (Japan and Korea)	10.7	10.5	10.8
Europe, Africa, the Middle East	13.3	13.5	14.4
USAs	10.5	9.9	9.7
Australia, New Zealand and South Pacific	5.4	7.6	6.4
Total	*N = 354 (100)*	*N = 304 (100)*	*N = 362 (100)*

4.2 Demographic characteristics of the sample for the airlines sector

The overall gender distribution for respondents in the airline sector was 52.5% male and 47.2% female while 0.3% answered that they were not sure. About 14.7% were aged between 16 and 25 years old, 32.5% between 26 and 35, 22.0% between 36 and 45, 21.5 between 46 and 55, 7.6% between 56 and 65, and 1.7 were aged 66 or above. Most of the respondents were married with kids (52.8%), with 33.6% being single, 12.1% married with no kids, 1.1% divorced or separated and 0.3% a widow or widower.

For the government sector, the majority of respondents were females (50.7%). The majority were aged 26–35 (30.3%) followed by those aged 36–45 (25.7%). In this sector, the majority of respondents were married with kids (49%) followed by singles (32.2%). Lastly, in the transport sector, the majority were males (58.3%). The majority were also in the age bracket of 26–35 (38.7%) followed by 36–45 (25.1%) and with the least 66 and above (1.4%). The majority were also married with kids (48.9%) followed by singles (35.9%).

In addition, most respondents were highly educated with an accumulated total of over 84% having some college or university education or above. In terms of occupation, most of the respondents were professionals (e.g., lecturer, doctor, accountant) at 38.4%, with the self-employed (e.g., businessman/woman) being the next largest group, followed by sub-professionals (semi-skilled worker, e.g., clerical officer) at 16.4%. In terms of reporting their household income, 35.1% of respondents reported their income in Renminbi (RMB) and 55.2% reporting in US$, with 12.1 reporting their income as RMB100,001 and above, and 13.8% reporting their income as US$100,001 and above.

Other relevant information relates to language with 44.9% responding in English, 7.1% in Cantonese and 48.0% in Putonghua. Also, in terms of the way the respondents were travelling, most were independent travellers (81.6%) whilst the rest (18.1%) were on all-inclusive package tours with a tour guide (0.3% did not answer this question).

4.3 Cluster and discriminant analysis

Cluster and discriminant analyses were conducted for the three tourism sectors of government, airlines and transport. This section reports and analyses the findings for each sector beginning with government, airlines and lastly the transport sector.

4.3.1 Government services

For the government sectors it was noted that visitors have no choice whether to use the police, customs or immigration. It was then logical to just apply the analysis to the Leisure and Cultural Services Department (LCSD) services that include museums, galleries, cultural events/programmes and public parks. This helped in drawing conclusions about the service attributes that perhaps would influence visitors' intention to re-use parks, cultural facilities, museums and events if they were to return to Hong Kong.

The results were firstly assessed for validity. Cluster analysis provided us with a 'fair' rating for the distribution of our clusters at 67.5% for value 1 representing those who answered 'yes' and 32.5% for value 2 for those who answered 'no'. The Tests of Equality of group mean table provided us with strong statistical evidence of significant differences between 15 out of 18 of our service attributes, with correspondingly low inter-correlations. The three statistically insignificant questions related to a good selection of exhibition items, cleanliness of toilets in public parks and the provision of snack stalls.

Box's M tests the null hypothesis that the covariance matrices do not differ between groups formed by the dependent. This test is expected not to be significant so that the null hypothesis that the groups do not differ can be retained. In this case, the log determinants appear similar and Box's M was 1.57 with F = 1.50 which is significant at $p < .001$.

Results in Table 2 show the Eigenvalue of 3.711 and canonical correlation of .888 which is interpreted as the proportion of variance explained (R^2). A canonical correlation of .888 suggests the model explains 78.85% of the variation in the grouping variable, i.e., those likely to re-use LCSD's services or not if they return to Hong Kong. Wilks' lambda confirms the significance of the discriminant function. This test was significant ($p < .001$) and indicated that 21.2% of the total variance remained not explained.

Table 2 Discriminant analysis summary

Discriminant function	*Eigenvalue*	*% of variance explained*	*Canonical correlation*	*Wilks's lambda*	*Chi-square*	*df*	*p*
1	3.711	100	0.888	0.212	42.622	3	0.000

The standardised canonical discriminant functions coefficients table provides an index of importance of predictors. The results from this test indicated that there are three aspects that are most influential in visitors' intention to re-use LCSD services; convenience of access to museums, convenience of access to public parks, and security and safety at the facilities/events. However, the structure matrix provides a better indication of issues of influence (with a value above 0.3), namely: clear and informative signage, pleasant environment of museums and galleries, and readily available information about museums and galleries and their locations.

The canonical discriminant function coefficients show the unstandardised coefficients (*b*) which are used to create the discriminant equation. In this case:

$$D = (.655 \times d2b) + (1.827 \times d10b) + (1.310 \times d17b) - 15.612$$

where *d2b* is convenient access to museums and galleries; *d10b* is convenient access to public parks and*d17b* is security and safety at the facilities/events. The group centroids results indicated that those who indicated intention to use the services again had a mean score of −1.192 while those who said 'no' produced a mean of 2.913 as shown in Table 3.

Prior probabilities for groups shows the proportional by chance accuracy rate which is computed by squaring and summing the proportion of cases in each group from the table of prior probabilities for groups as ($0.710^2 + 0.290^2$ ′ 1.25 = 0.73525). This means that our cross-validated accuracy should be 73.52%.

Table 3 Discriminant function loadings – functions at group centroids

Functions at group centroids	
Two step cluster number	*Function*
	1
1	–1.192
2	2.913

Note: Unstandardised canonical discriminant functions evaluated at group means.

The classification results reveal that 100% of respondents were classified correctly into those who intend to use the leisure and cultural services or those who will not. The overall predictive accuracy of the discriminant function also known as the 'hit ratio' was perfect at 100%.

4.3.2 Airlines

For airlines, the two-step cluster analysis provided the distribution of 63.1%, for value 1 representing those who answered 'yes' and 36.9% for value 2 for those who answered 'no'. The tests of equality of group mean revealed a strong statistical evidence of significant differences for all the ten service attributes, with correspondingly low inter-correlations.

Box's M tests the null hypothesis that the covariance matrices do not differ between groups formed by the dependent. This test is expected not to be significant so that the null hypothesis that the groups do not differ can be retained. In this case, the log determinants appear similar and Box's *M* was 16.91 with $F = 1.62$ which is not significant at $p > .001$

Table 4 Discriminant analysis summary

Discriminant function	*Eigenvalue*	*% of variance explained*	*Canonical correlation*	*Wilks's lambda*	*Chi-square*	*df*	*p*
1	1.881	100	0.808	0.347	113.208	4	0.000

The eigenvalues provided a canonical correlation of .808 which is interpreted as the proportion of variance explained (R^2). This value suggests that the model explains 65.3% of the variation in the grouping variable, i.e., those likely to re-use the airline services or not if they return to Hong Kong. Wilks' Lambda confirms the significance of the discriminant function. This test was significant ($p < .000$) and indicated that 34.7 % of the total variance remained not explained.

The standardised canonical discriminant functions coefficients table provides an index of importance of predictors. However, the structure matrix provides a better indication of issues of influence (with a value above 0.3), namely: proper attitude of frontline staff (polite, patient and attentive), efficient check-in and baggage handling services of the airline, promptness and professionalism in handling complaints, and clean and comfortable interiors and seats of aircraft.

The canonical discriminant function coefficients provided the unstandardised coefficients (*b*) which are used to create the discriminant equation. In the case of airlines,

$$D = (.565 \times a1b) + (1.043 \times a7b) + (.934 \times a9b) + (.814 \times a10b) - 13.871$$

where $a1b$ is clean and comfortable interiors and seats of aircraft, $a7b$ is efficient check-in and baggage handling services of the airline), $a9b$ is proper attitude of frontline staff (polite, patient and attentive) and $a10b$is promptness and professionalism in handling complaints.

A further interpretation of discriminant analysis using functions at group centroids results as in Table 5indicated that those who indicated intention to use the services again had a mean score of 1.776 while those who said 'no' produced a mean of –1.040. Under this test, all cases with scores near to a centroid are predicted as belonging to that group.

Table 5 Discriminant function loadings – functions at group centroids

Functions at group centroids	
Two step cluster number	*Function*
	1
1	1.776
2	–1.040

Note: Unstandardised canonical discriminant functions evaluated at group means.

Prior probabilities for groups test shows the proportional by chance accuracy rate which is computed by squaring and summing the proportion of cases in each group from the table of prior probabilities for groups. The proportional by chance accuracy rate for the airline sector is $(0.369^2 + 0.631^2) = .534$.

This means that a 25% increase over this would require that our cross-validated accuracy be 1.25* .534% = 66.8%. The classification function coefficient using Fisher's linear discriminant function reveals that 90.2% of respondents were classified correctly into those who intend to use the airline services when they return to Hong Kong or those who will not at 95.7%. The overall predictive accuracy of the discriminant function, also known as the 'hit ratio', showed 93.7% of original grouped cases were correctly classified.

4.3.3 Public transport

For railway services, the results show significant differences of the means of two groups (i.e., 'will use the railway services' or 'will not use the railway services') as all independent variables produce high value F's. It is also reflected from the *pooled within-group matrices* that the inter-correlations of the independent variables are low. While the log determinants of two groups appear similar, box's

M indicates that the assumption of equality of covariance matrices is violated (i.e., Box's M is 14.676 with F = 2.249 that is significant at p < .000). However, given the large sample, this problem is not regarded as serious.

Table 6 Discriminant analysis summary

Discriminant function	*Eigenvalue*	*% of variance explained*	*Canonical correlation*	*Wilks's lambda*	*Chi-square*	*df*	*p*
1	2.355	100	0.838	0.298	47.808	3	0.00(

A significant discriminant function was derived, with eigenvalues 2.355 and a canonical correlation of 0.838 suggesting that the model explains 70.2% of the variance. The proportion of total variability not explained is 29.8% as indicated by Wilks' lambda. Group membership is significantly affected by 3 of the 12 attributes. 'Clear announcements at the stations and on the trains' turns out to be the strongest predictor contributing 70% of the Function 1 structure, while 'feeling safe when using the services' and an 'efficient and easy-to-follow ticketing system' come second and third respectively. The discriminant function is:

$$D = (1.164 \times t3b) + (0.663 \times t4b) + (1.168 \times t7b) - 12.728$$

where *t3b* refers to clear announcements at the stations and on the trains, *t4b* refers to an efficient and easy-to-follow ticketing system, and *t7b* refers to feeling safe when using the services.

According to the figures of *group centroids* in Table 7, those who will use the railway services again have a mean of 1.397 while those who will not use the railway services produce a mean of –1.607. The classification results reveal that 93% of respondents were classified correctly into 'will use the railway services' or 'will not use the railway services'. As for the overall predictive accuracy of the discriminant function, 'will use the railway services' has an accuracy of 91.3% while 'will not use the railway services' has 90%.

Table 7 Discriminant function loadings – functions at group centroids

Cluster membership	*Function 1*
1	1.397
2	–1.607

Note: Unstandardised canonical discriminant functions evaluated at group means.

Discriminant analysis for bus services results show significant differences of the means of two groups (i.e., 'will use the bus services' or 'will not use the bus services') as all independent variables produce high value F's. It is also reflected from the *pooled within-group matrices* that the inter-correlations of the independent variables are low. The log determinants of two groups appear similar and box's M is 14.675 with F = 2.249, which is insignificant as p > .000, and it suggests that two groups are significantly different.

Table 8 Discriminant analysis summary

Discriminant function	*Eigenvalue*	*% of variance explained*	*Canonical correlation*	*Wilks's lambda*	*Chi-square*	*df*	*p*
1	2.263	100	0.833	0.306	69.187	3	0.000

A significant discriminant function was derived, with Eigenvalues 2.263 and a canonical correlation of 0.833 suggesting that the model explains 69.4% of the variance. The proportion of total variability not explained is 30.6% as indicated by Wilks' lambda. Group membership is significantly affected by three of the ten attributes. 'Proper attitude of bus driver (polite and patient)' is the strongest predictor contributing 77.1% of the Function 1 structure, followed by 'feeling safe on board' and 'well-maintained and clean buses'. The discriminant function is:

$$D = (0.731 \times b1b) + (0.678 \times b6b) + (1.087 \times b9b) - 9.467$$

where *b1b* refers to well maintained and clean buses, *b6b* refers to feeling safe on board, and *b9b* refers to proper attitude of bus driver – polite and patient.

According to the figures of *group centroids* in Table 9, those who will use the bus services again have a mean of 2.314 while those who will not use the bus services produce a mean of –0.947. The classification results reveal that 95.2% of respondents were classified correctly into 'will use the bus services' or 'will not use the bus services'. As for the overall predictive accuracy of the discriminant function, 'will use the bus services' has an accuracy of 88.9% while 'will not use the bus services' has 97.7%.

Table 9 Discriminant function loadings – functions at group centroids

Cluster membership	*Function 1*
1	2.314
2	–.947

Note: Unstandardised canonical discriminant functions evaluated at group means.

Lastly, discriminant analysis for taxi results show significant differences of the means of two groups (i.e., 'will use the taxi services' or 'will not use the taxi services') as all independent variables produce high value F's. It is also reflected from the *pooled within-group matrices* that the inter-correlations of the independent variables are low. The log determinants of two groups appear similar and box's M is 0.783 with F = 0.773, which is insignificant as p > .000, and it suggests that two groups are significantly different.

Table 10 Discriminant analysis summary

Discriminant function	*Eigenvalue*	*% of Variance explained*	*Canonical correlation*	*Wilks's lambda*	*Chi-square*	*df*	*p*
1	3.893	100	0.892	0.204	128.607	6	0.000

A significant discriminant function was derived, with eigenvalues 3.893 and a canonical correlation of 0.892 suggesting that the model explains 79.6% of the variance. The proportion of total variability not explained is 20.4% as indicated by Wilks' lambda. Group membership is significantly affected by six of the eight attributes. 'Help with loading and unloading', 'honesty of taxi drivers', and 'language and communication skills of taxi drivers' are the three most important predictors as they have the largest coefficients of the Function 1 structure. The discriminant function is:

$$D = (0.527 \times b12b) + (0.489 \times b13b) + (0.628 \times b14b) + (1.034 \times b15b) + (0.718 \times b16b) + (1.107 \times b18b) - 18.841$$

where *b*12*b* refers to clear and accurate information about fares and destinations, *b*13*b* refers to the tidiness of taxi drivers, *b*14*b* refers to the language and communication skills of taxi drivers, *b*15*b* refers to the honesty of taxi drivers, *b*16*b* refers to feeling safe on board, and *b*18*b* refers to help with loading and unloading.

According to the figures of *group centroids* in Table 11, those who will use the taxi services again have a mean of 0.881 while those who will not use the taxi services produce a mean of 1.159. The classification results reveal that 98.8% of respondents were classified correctly into 'will use the taxi services' or 'will not use the taxi services'. As for the overall predictive accuracy of the discriminant function, 'will use the taxi services' has an accuracy of 98% while 'will not use the taxi services' has 100%.

Table 11 Discriminant function loadings – functions at group centroids

Cluster membership	*Function 1*
1	0.881
2	1.159

Note: Unstandardised canonical discriminant functions evaluated at group means.

4.4 Importance performance analysis for each sector

In order to establish the importance of each of the service quality attributes in determining each sector's performance, IPA was conducted on the airlines, transport, and government sectors. This section shows the results.

From Figure 2, it can be seen that the airline sector is performing extremely well in the area of providing clean and comfortable interiors and seats of aircraft, prompt service upon request by flight attendants, efficient handling of reservation, cancellation and confirmation requests, efficient check-in and baggage handling services of the airline, language and communication skills, proper attitude of frontline staff in terms of politeness, patience and attentiveness, and promptness and professionalism in handling complaints. According to the IPA model, all of these lie in quadrant B which shows that customers were satisfied with the services therein and they attached greater importance to them. Therefore, the airlines in Hong Kong should keep up the good work with regard to these service attributes.

In general, the airline sector in Hong Kong seems to have met the customer expectations. The only problem area identified (quadrant C; poor performance) is, fortunately, of less importance. Interestingly, there is virtually no 'overkill'; also, there is nothing important overlooked. This is in line with evidence from elsewhere: many independent surveys reveal that Asian and South East Asian Airlines meet customer expectations better than their counterparts based in other parts of the world. For instance, out of the top ten, SKYTRAX airline of the year 2011 award winners, seven were from Asia.

4.4.1 Public transport

From Figure 3, Hong Kong railways have performed very well in many aspects. Visitors are satisfied with the journey provided especially because of: feeling safe, punctuality and reliability of service, clear and accurate directional signage and location maps, and clear announcements. Besides this, all these aspects are considered important to the visitors. The findings imply that Hong Kong railways should keep up the good work in all these services attributes.

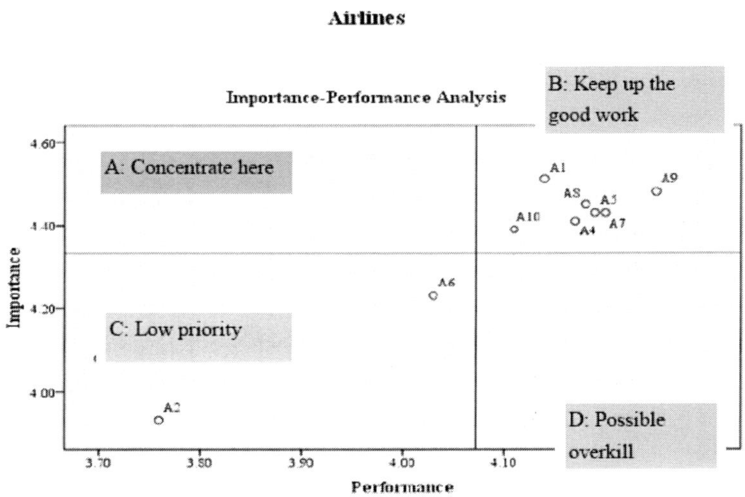

Notes: *Key:* A1: clean and comfortable interiors and seats of aircraft
A2: in-flight entertainment facilities
A3: quality and variety of in-flight meals and drinks
A4: prompt service upon request by flight attendants
A5: efficient handling of reservation, cancellation and confirmation requests
A6: able to arrange preferred seat at check-in
A7: efficient check-in and baggage handling services
R8: language and communication skills of staff
R9: proper attitude of frontline staff (polite, patient, attentive)
R10: promptness and professionalism in handling complaints.

Figure 2 The airline sector: importance – performance analysis (see online version for colours)

However, Hong Kong railways have not done so well in terms of the provision of toilets at stations, and the language and communication skills of staff, although these two aspects are also deemed important to visitors. Hong Kong railways thus should focus more on improving its service within these two areas.

On the other hand, the services provided by the bus companies are not as good as the railway services. As seen from the IPA results, the bus companies are able to maintain well-cleaned buses, to be punctual and reliable, and to provide a safe feeling, and these aspects have been viewed as important to visitors. However, in another important area, i.e., clear and accurate signs and route information at bus stops, the performance needs to be improved.

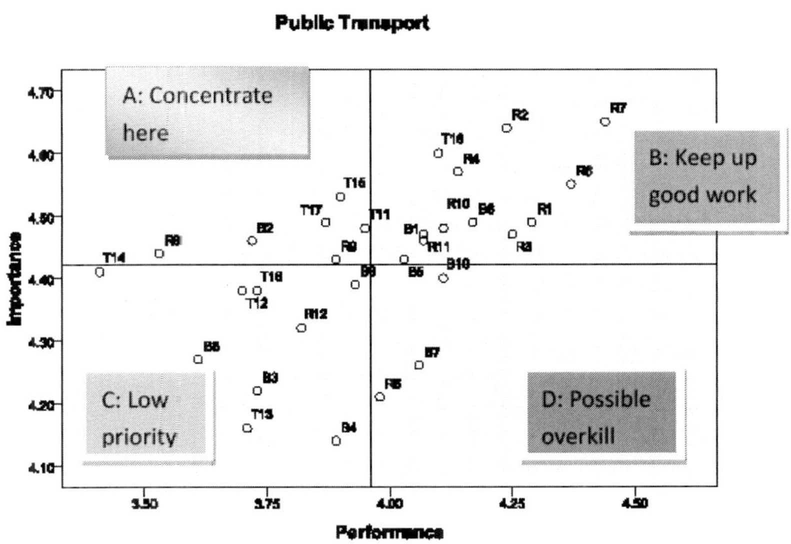

Notes: *Key:* 'R': Hong Kong railways
'B': Hong Kong franchised buses
'T': Hong Kong taxis
R1: clean and pleasant compartments/platform
R2: clear and accurate directional signage and location maps
R3: clear announcement
R4: efficient and easy-to-follow ticketing system
R5: provision of tourist transport passes
R6: punctuality and reliability of service
R7: feeling safe
R8: provision of toilets at stations

R9: language and communication skills of staff
R10: proper attitude of staff
R11: efficient handling of enquiry
R12: promptness and professionalism in handling complaints
B1: well maintained and clean buses
B2: clear and accurate signs and route information at bus stops
B3: provision of tourist buses
B4: provision of tourist transport passes
B5: punctuality and reliability of service
B6: feeling safe on board
B7: tidiness of bus drivers
B8: language and communication skills of bus drivers
B9: proper attitude of bus driver
B10: appropriate operating hours
T11: well maintained and clean taxis
T12: clear and accurate information
T13: tidiness of taxi drivers
T14: language and communication skills of taxi drivers
T15: honesty of taxi drivers
T16: feeling safe on board
T17: professional attitude of taxi drivers
T18: help with loading and unloading.

Figure 3 The public transport sector: IPA (see online version for colours)

As for taxi services, it appears to be performing least well in comparison with the other two modes of public transport based upon the IPA findings as shown in Figure 3. Taxi services do well in the provision of safe feeling, where the taxi service should keep up the good work in future. However, the areas of the honesty of taxi drivers as well as their professional attitude fall short of the visitors' satisfaction and require additional attention for further improvement. Along with the aspect of maintaining clean taxis that also needs to be improved, these areas are all perceived to be important to visitors.

4.4.2 Government service: police

A more detailed analysis of each sub-sector reveals areas where each service could make improvements relative to its overall performance. The Hong Kong Police score well for giving directions to local amenities (p1), projecting a feeling of safety by their presence (p5) and having a polite and patient attitude (p6). If there is such a thing as being 'too' efficient then, the Hong Kong police may be guilty as charged in dealing with enquiries efficiently (p6), but this is hardly something one would wish to change. Where they score relatively poorly in quadrant C – in terms of visibility in public (p4) – does not seem to be a major issue for respondents. It is in terms of their language and communication skills (p3) where the Hong Kong Police should be concentrating (quadrant A). It is worth going into more depth on this particular question to ascertain which particular groups of respondents felt most strongly about this and where they were from in order to determine which language skills need brushing up.

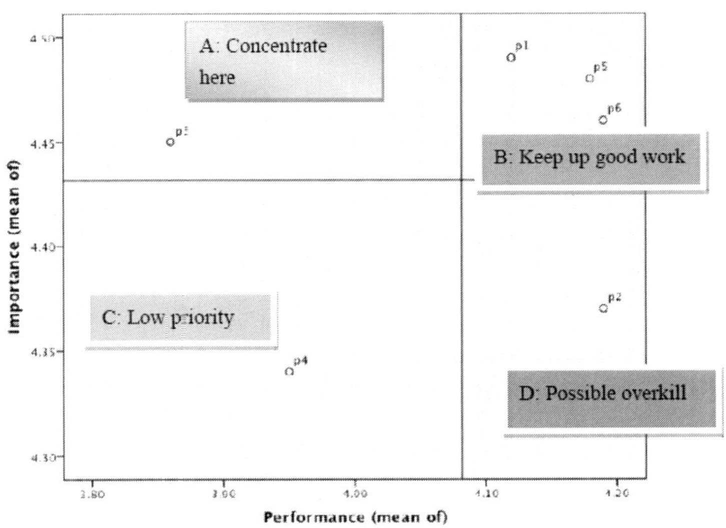

Notes: *Key:* p1: police officers have good knowledge of directions and local amenities when asked by visitors
p2: police officers deal with enquiry efficiently
p3: language and communication skills of police officers
p4: visibility of police officers in public
p5: feeling safe with police presence
p6: proper attitude of police officers (polite and patient).

Figure 4 The police sector: importance – performance analysis (see online version for colours)

The above IPA, however, is intriguing to some extent: respondents express serious concerns about the communication skills of the police officers. At the same time, they seem to play down the efficiency with which the police investigate inquiries. It is difficult to understand why the efficient way police deals with inquiries is not important but at the same time their communication skills are important. Perhaps the language incompetence of the police officers makes the fruits of their labour difficult to be appreciated by the tourists. Alternatively, this situation might be reflective of the nature of the tourist psyche involving playfulness and lack of depth (Urry, 1992).

4.4.3 Immigration

The Hong Kong immigration service is well regarded by the sample tourists in terms of clear signage to the clearance counters (i1), clear instructions on procedures at the border (i3), the language and communication skills of the immigration officers (i5), as well as their proper manner (i6). There are no major areas requiring great focus to improve but they score relatively poorly for the queuing environment (i2) and queuing time – even though these are given a relatively low priority by respondents.

4.4.4 Customs

The record for Hong Kong customs is somewhat more mixed. They have only two service attributes of which they can be proud in quadrant B – the language and communications skills of their officers (c5) and their attitude (c7). They may be overly concerned with courtesy at security checking (c6) as they score highly for this but it is regarded as of relatively low importance by the tourist sample. The environment of the customs area (c2) is on the cusp of being overdone and of low priority. Also regarded as of low priority by respondents is the clarity of the instructions about customs regulations (c3) and clear signage (c1), although the latter is tending towards being something that there should be more focus on. The main area for the customs service to focus on is making the queuing time at customs for all types of crossings less than 15 minutes (c4), although this also tended to be regard as not a relatively high priority.

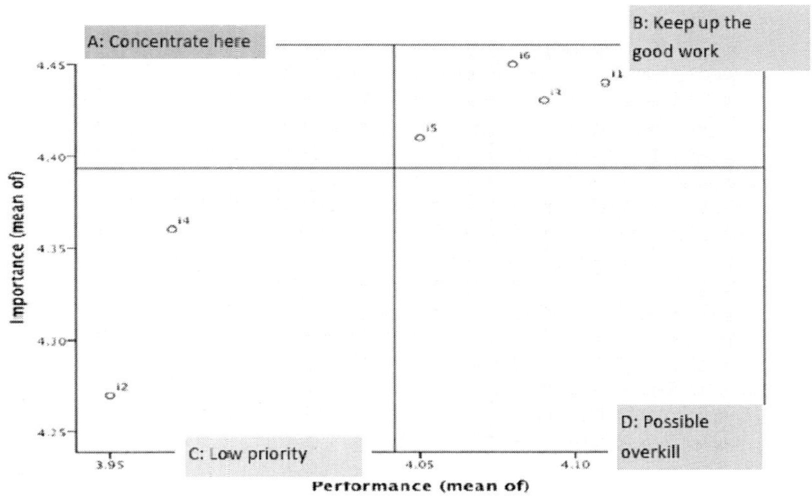

Notes: *Key*: i1: clear signage to clearance counters for visitors
i2: pleasant environment of the queuing area
i3: clear instruction of immigration procedures at the border
i4: less than 15 minutes queuing time for the clearance (air) or less than 30 minutes queuing time for clearance (by sea/land)
i5: language and communication skills of immigration officers
i6: proper attitude of immigration of officers (polite and patient).

Figure 5 The immigration services sector: importance – performance analysis (see online version for colours)

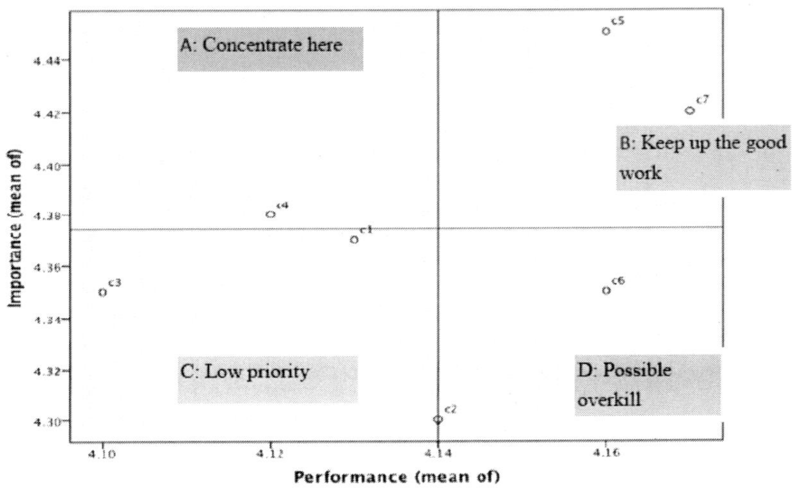

Notes: *Key:* c1: clear signage to the customs area for visitors
c2: pleasant environment of the customs area for visitors
c3: clear instruction of customs regulations
c4: less than 15 minutes of queuing time at the customs
c5: language and communication skills of customs officers
c6: thorough yet courteous security checking at the counter
c7: proper attitude of customs officers (polite and patient).

Figure 6 IPA for customs (see online version for colours)

4.4.5 LCSD services

This section of the survey covered a wide range of services including museums/galleries (d1–d5), cultural events/programmes (d6–d8), public parks (d9–d13) and attributes related to all of those services. The museums sector is excelling in providing clear and multi-lingual interpretation of exhibition items (d5), as well as providing pleasant environments (d1) and convenient access (d2) – although these last two were not very highly rated in terms of importance. Museums and galleries do not appear to have any serious issues to concentrate on but should be wary of their performance in the areas of availability of information about what they offer (d4) and having a good selection of exhibition items (d3). An area of improvement as a whole could be cultural events and programmes.

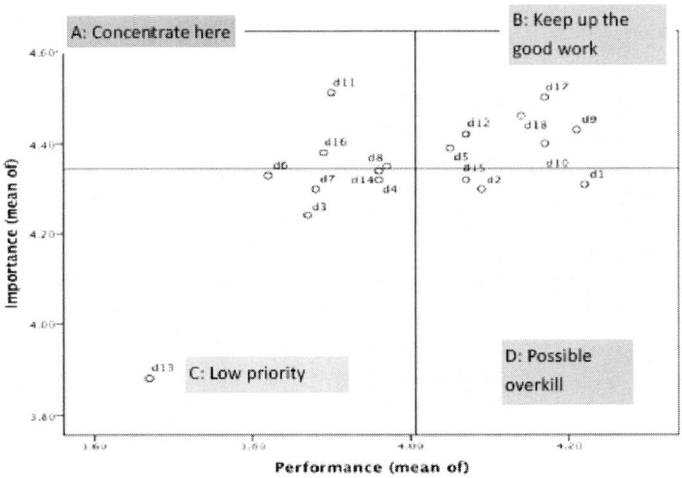

Notes: Key: d1: pleasant environment of museums and galleries
d2: convenient access to museums and galleries
d3: good selection of exhibition items
d4: readily available information about museums, galleries and their location
d5: provision of clear and multi-lingual interpretation of exhibition items
d6: effective promotion of cultural events and programmes
d7: variety of cultural events and programmes
d8: quality of cultural events and programmes
d9: clean and pleasant environment of public parks
d10: convenient access to public parks
d11: cleanliness of toilets in public
d12: clear and informative signage within public parks
d13: provision of food/snack stalls
d14: staff's knowledge about the facilities/events/locations
d15: tidiness of staffd16: Language and communication skills of staff
d17: security and safety at the facilities/events
d18: proper attitude of frontline officers (polite, patient and attentive).

Figure 7 IPA for leisure and culture services (see online version for colours)

All three service attributes found themselves hovering around the cusp of quadrant C and A, and it is clearly an area in which Hong Kong could improve as a whole. Hong Kong's parks are popular amongst tourists for their cleanliness and nice environment (d9), convenient access (d10) and clear signage (d12). Even though the provision of snack stalls (d13) is regarded as relatively poor, it not regarded as a particular priority by the respondents. However, it would appear that LCSD should concentrate on maintaining the cleanliness of public toilets in public parks (d11).

Overall, the LCSD are performing well in providing security and safety at venues (d17), their staff should be congratulated on their polite, patient and attentive attitude (d18), as well as their tidiness (d15), although this last attribute was not regarded highly important. Staff knowledge about facilities/events/locations was relatively poor and though in quadrant C tends towards being something that may require attention in the future. Again, as for the police, an area of concentration for further training for the LCSD is the language and communication skills of the staff and it can again be analysed which language they need training in by looking more closely at the respondents who rated this question as important and were relatively dissatisfied with the service they received.

5 CONCLUDING REMARKS

This study has attempted to investigate the perceptions of quality of service offered by the tourism and related industries for visitors to Hong Kong to ascertain how these industries are performing in providing quality service to match up with the tourists' expectations and provide useful quantitative information for service quality improvement to the relevant sectors. Methods used were cluster and discriminant analysis to determine the most significant factors in deciding whether respondents would re-use the service or not, and IPA to determine where these sectors should focus resources to provide the most needed improvement based on the gap between visitor expectation of service and satisfaction.

The results from the present study show that for the LCSD the factors that are most persuasive in getting visitors to re-use their services are convenience of access to museums, convenience of access to public parks, and security and safety at the facilities/events. For airlines, the factors of persuasion were proper attitude of frontline staff (polite, patient and attentive), efficient check-in and baggage handling services of the airline, promptness and professionalism in handling complaints, and clean and comfortable interiors and seats of aircraft. For railways, clear announcements at the stations and on the trains, feeling safe when using the services, and efficient and easy-to-follow ticketing system were most influential in eliciting a likely re-use of the service, whilst for buses it was the proper attitude of bus drivers (polite and patient), feeling safe on board, and well-maintained and clean buses. For taxis, the most effective predictors of re-use were help with loading and unloading, the honesty of taxi drivers, and the language and communication skills of taxi drivers.

The analysis of the importance performance figures for each sector indicates clearly the areas in which each sector needs to concentrate (quadrant A). Simply using IPA studies alone, however, would still provide only part of the picture and it is believed that by using them in combination with cluster and discriminant analysis these sectors would yield more reliable results. The sectors then would be able to adopt a more cost-effective 'two-pronged' approach of focusing resources not only on improving the negative performance aspects of their services (as revealed by IPA) but also the positive factors most likely to bring re-purchase (as revealed by cluster and discriminant analysis). Unfortunately, proving repurchase appears to be difficult as the data that follow up on these particular respondents' future behaviour in this regard are not available (Edward and George, 2008).

An interesting observation from the study is that safety for airlines did not even make the list whereas it shows strongly in the other modes of transportation (we thank the anonymous reviewer for pointing out this 'anomaly'). The technological advancements and generally excellent airline infrastructure in Asia could be a reason for this finding. It could be that safety is assumed by passengers 'by default' and hence is not a concern.

Another observation worthy of further consideration is that the analysis found no association between gender and intention to use services for airlines and transport. In contrast, a significant association was found between gender and intention to use leisure and cultural services. The transportation sector in general provides services aimed at both the genders whereas leisure and cultural services are generally gender insensitive. When certain services are gender insensitive, different genders might respond to such services differently. An alternate explanation is that transportation is a support service or augmented product whereas leisure and cultural services are the core products (attractions) for the consumers. Tourists to Hong Kong might be focusing entirely on the quality of the attractions provided the quality of support services exceeds a critical minimum. The IPA performed by us probably agrees with this explanation.

While many primary relationships in the field of consumer behaviour are pretty generic and are already well known, the ramifications of them for particular contexts are not well evident. In the present paper, we investigated one such specific context. Among other things, the survey provided data on the perceptions of value for money for each sector. This study suggests further analysis is required of the role of perceived value for money as a moderating factor impacting customer loyalty within the conceptual model (Figure 1). This examination will provide deeper meaning and explanation to our observation that customer loyalty cannot focus solely on performance and/or service quality.

6 LIMITATIONS AND RECOMMENDATION FOR FUTURE RESEARCH

Like many studies that use convenience sampling and have a limitation of generalisability of the study findings, this study does not claim generalisability. This is because it is extremely difficult to obtain a sampling frame to enable random sampling methods like systematic sampling. However most research that uses the intercepting of respondents at entry or exit points has been able to obtain rich data that is used to infer to the general population.

Secondly, the sample sizes used are relatively small compared to the total number of visitors that enter or leave Hong Kong using the terminals we used to collect data from. However a sample size above 200 has been found to be good enough to enable conducting of inferential statistical methods (Hair et al., 2010). To improve on the sample size, future studies can consider sector by sector and collect larger sample sizes and compare the sectors after, unlike the concurrent data collection carried out in this study.

Finally, it is not that other sectors are any less important or the exclusion of a particular issue from the scope of this paper does not necessarily mean that it lacks merit. We must add that this study is part of a larger research agenda that aims to reassess the potential of Hong Kong tourism. Problems and prospects related to some of the other sectors such as attractions and accommodations that co-determine the competence of Hong Kong tourism are currently being analysed.

AUTHOR DETAILS

Eddy K. Tukamushaba is a PhD research student at the Hong Kong Polytechnic University in the School of Hotel and Tourism

Management in the area of destination branding. He is a Lecturer in the Department of Leisure and Hospitality in the Faculty of Marketing and Hospitality Management at Makerere University Business School, Kampala, Uganda. His publications have appeared in different journals like; *Journal of International Entrepreneurship, Journal of Tourism, Journal of Teaching in Travel and Tourism, International Journal of Tourism Sciences, International Journal of Market Research*, and *Journal of China Tourism Research*. He was the winner of Emerald – ALCS Research Fund Award 2008 Public Sector Management in Africa and is a member of International Academy for African Business Development (IAABD).

Babu P. George is an Associate Professor in the Department of Business Administration of Alaska Pacific University. He is an interdisciplinary researcher mainly focusing on consumption in the service industries. He has published more than fifty research papers in peer reviewed academic journals. Currently, he is the Editor-in-Chief of *International Journal of Qualitative Research in Services* and also serves on the editorial/review boards of many journals.

REFERENCES

1 Ajzen, I. , Driver, B.L. (1992). 'Application of theory of planned behavior to leisure choice'. Journal of Leisure Research. 24, 3, 207-224

2 Akama, J.S. , Kieti, D.M. (2003). 'Measuring tourist satisfaction with Kenya's wildlife safari: a case study of Tsavo West National Park'. Tourism Management. 24, 1, 73-81

3 Armstrong, R. , Mok, C. , Go, F.M. , Chan, A. (1997). 'The importance of cross-cultural expectations in the measurement of service quality perceptions in the hotel industry'. International Journal of Hospitality Management. 16, 2, 181-190

4 Atilgan, E. , Akinci, S. , Aksoy, S. (2003). 'Mapping service quality in the tourism industry'. Managing Service Quality. 13, 5, 412-422

5 Boulding, W. , Kalra, A. , Staelin, R. , Zeithmal, V.A. (1993). 'A dynamic process model of service quality: from expectations to intentions'. Journal of Marketing Research. 30, 1, 7-27

6 Brown, T.J. , Churchill, G.A. , Peter, J.P. (1993). 'Improving the measurement of service quality'. Journal of Retailing. 66, 1, 127-139

7 Chew, J. (1987). 'Transport and tourism in the year 2000'. Tourism Management. 8, 2, 83-85

8 Choi, T.Y. , Chu, R. (2001). 'Determinants of hotel guests' satisfaction and repeat patronage in the Hong Kong hotel industry'. International Journal of Hospitality Management. 20, 3, 277-297

9 Chu, R.K.S., Choi, T. (2000). 'An importance-performance analysis of hotel selection factors in the Hong Kong hotel industry: a comparison of business and leisure travelers'. Tourism Management. 21, 4, 363-377

10 Churchill, G.A. , Surprenant, C. (1982). 'An investigation into the determinants of customer satisfaction'.Journal of Marketing Research. 19, 4, 491-504

11 Cronbach, L.J. (1951). 'Coefficient alpha and the internal structure of tests'. Psychometrika. 16, 8, 297-334

12 Crouch, G. , Ritchie, J. (1999). 'Tourism, competitiveness, and societal prosperity'. Journal of Business Research. 44, 3, 137-152

13 Dabholkar, P.A. , Thorpe, D.I. (1994). 'Does customer satisfaction predict shopper intentions?'. Journal of Consumer Satisfaction, Dissatisfaction and Complaining Behaviour. 7, 161-171

14 Deng, W. (2007). 'Using a revised importance-performance analysis approach: the case of Taiwanese hot springs tourism'. Tourism Management. 28, 5, 1274-1284

15 Dube, L. , Renagham, L.M. , Miller, J.M. (1994). 'Measuring customer satisfaction for strategic management'.Cornell Hotel and Restaurant Administration Quarterly. 35, 1, 39-48

16 Edward, M. , George, B.P. (2008). 'Tourism development in the State of Kerala, India: a study of destination attractiveness'. European Journal of Tourism Research. 1, 1, 16-38

17 Enright, M.J., Newton, J. (2004). 'Tourism destination competitiveness: a quantitative approach'. Tourism Management. 25, 6, 777-788

18 Fornell, C. (1992). 'A national customer satisfaction barometer: the Swedish experience'. Journal of Marketing. 56, 1, 6-21

19 George, B.P. (2005). 'Measuring tourist attachment to holidays: some preliminary results'. Tourism: An International Interdisciplinary Journal. 52, 3, 229-246

20 Gilbert, D. , Wong, R.K.C. (2003). 'Passenger expectations and airline services: a Hong Kong based study'.Tourism Management. 24, 5, 519-532

21 Gomezelj, D.O. , Mihalic, T. (2008). 'Destination competitiveness-applying different models the case of Slovenia'. Tourism Management. 29, 2, 294-307

22 Gotlieb, J.B. , Grewal, D. , Brown, S.W. (1994). 'Consumer satisfaction and perceived quality: complementary or divergent constructs?'. Journal of Applied Psychology. 79, 6, 875-885

23 Gronroos, C. , Swartz, T.A. , Bowen, D.A. , Brown, S.W. (1993). 'Toward a third phase in service quality research: challenges and

future directions'. Advances in Services Marketing and Management. 2, Greenwich, CT:JAI Press , 49-64

24 Gupta, S. , Lehmann, D. , Stuart, J. (2004). 'Valuing customer'. Journal of Marketing Research. 41, 1, 7-18

25 Haemoon, O. (2001). 'Revisiting importance-performance analysis'. Tourism Management. 22, 6, 617-627

26 Hair, J.F. , Black, W.C. , Babin, B.J. , Anderson, R.E. (2010). Multivariate Data Analysis. A Global Perspective. 7th ed., Upper Saddle River, New Jersey:Pearson Educatio, Inc.

27 Hansen, E. , Bush, R.J. (1999). 'Understanding customer quality requirements: model and application'.Industrial Marketing Management. 28, 2, 119-130

28 Howat, G. , Absher, J. , Crilley, G. , Milne, I. (1996). 'Measuring customer service quality in sports and leisure centers'. Managing Leisure. 1, 2, 77-89

29 Hui, T.K. , Wan, D. , Ho, A. (2007). 'Tourists' satisfaction, recommendation and revisiting Singapore'. Tourism Management. 28, 4, 965-975

30 Hsieh, L.F. , Lin, L.H. , Lin, Y.Y. (2008). 'A service quality measurement architecture for hot spring hotels in Taiwan'. Tourism Management. 29, 3, 429-438

31 Jin, X. , Qi, S.S. , Chiang, C.H. (2008). 'A cross-section evaluation of service quality in the tourism industry of Hong Kong: importance vs. performance'. Journal of China Tourism Research. 4, 3/4, 319-335

32 Johnston, R. (1995). 'The determinants of service quality: satisfiers and dissatisfiers'. International Journal of Service Industry Management. 6, 2, 53-71

33 Johnston, R. , Lyth, D. (1998). 'Service quality: Integrating customer expectations and operational capability'.in Proceedings of the QUIS Symposium. Sweden:University of Karlstad

34 Keaveney, S.M. (1995). 'Customer switching behavior in service industries: an exploratory study'. Journal of Marketing. 59, 2, 71-82

35 Khadaroo, J. , Seetanah, B. (2008). 'The role of transport infrastructure in international tourism development: a gravity model approach'. Tourism Management. 29, 5, 831-1052

36 Kim, Y.K. , Lee, H.R. (2011). 'Customer satisfaction using low cost carriers'. Tourism Management. 32, 2,235-243

37 Kim, W.G. , Ng, C.Y.N. , Kim, Y.S. (2009). 'Influence of institutional DINESERV on customer satisfaction return intention ad word-of-mouth'. International Journal of Hospitality Management. 28, 1, 10-17

38 Lacher, R.G. , Harrill, R. (2010). 'Going beyond sun, sand, and surf? An importance-performance analysis of activities in a 3S resort destination'. E-Review of Tourism Research. 8, 4, 57-68

39 Lee, G. , Lee, C. (2009). 'Cross-cultural comparison of the image of Guam perceived by Korean and Japanese leisure travelers: importance-performance analysis'. Tourism Management. 30, 6, 922-931

40 Lee, H. , Lee, Y. , Yoou, D. (2000). 'The determinants of perceived service quality and its relationship with satisfaction'. Journal of Service Marketing. 4, 2, 118-129

41 Martilla, J.A. , James, J.C. 'Importance-performance analysis'. Journal of Marketing. 1977, 01, 41, 1, 77-79

42 Nunnaly, J. (1978). Psychometric Theory. New York:McGraw-Hill

43 Oliver, R.L. (1980). 'A cognitive model of the antecedents and consequences of satisfaction decisions'.Journal of Marketing Research. 17, 4, 460-469

44 Oliver, R.L. , Swan, J.E. (1989). 'Equity and disconfirmation perceptions as influences on merchant and product satisfaction'. Journal of Consumer Research. 16, 3, 372-383

45 Parasuraman, A. , Zeithaml, V. , Berry, L. (1985). 'A conceptual model of service quality and its implications for future research'. Journal of Marketing. 49, 4, 41-50

46 Parasuraman, A. , Zeithaml, V. , Berry, L. (1988). 'SERVQUAL: a multiple-item scale for measuring consumer perceptions of service quality'. Journal of Retailing. 64, 1, 12-40

47 Parasuraman, A. , Zeithaml, V. , Berry, L. (1990). Developing Quality Service: Balancing Customer Perceptions and Expectations. New York:Free press

48 Park, J.W. , Robertson, R. , Wu, C.L. (2004). 'The effects of airline service quality on passenger's behavioral intentions: a Korean case study'. Journal of Air Transport Management. 10, 6, 435-439

49 Petrick, J.F. (2004). 'The roles of quality, value, and satisfaction in predicting cruise passengers' behavioural intentions'. Journal of Travel Research. 42, 4, 397-407

50 Poon, A. (1993). Tourism, Technology, and Competitive Strategies. Willingford, UK:CAB International

51 Prideaux, B. (2000). 'The role of the transport system in destination development'. Tourism Management. 21, 1, 53-63

52 Qu, H. , Sit, C.Y. (2007). 'Hotel service quality in Hong Kong: an importance and performance analysis'.International Journal of Hospitality and Tourism Administration. 8, 3, 49-72

53 Ritchie, J. , Crouch, G. (2003). The Competitive Destination: A Sustainable Tourism Perspective. New York:CAB International

54 Saleh, F. , Ryan, C. (1992). 'Client perceptions of hotels: a multi-attribute approach'. Tourism Management.13, 1, 163-168

55 Song, Z. , Kong, H. , Chen, Y. (2008). 'Tourist satisfaction, perceived service value and repurchase intentions: the case of Hong Kong's airline, hotel and restaurant services'. Journal of China Tourism Research. 4, 3/4,336-364

56 Song, H.Y. , Gang, L. , van der Veer, R. , Chen, J.L. (2011). 'Assessing Mainland Chinese tourists' satisfaction with Hong Kong using tourist satisfaction index'. International Journal of Tourism Research. 13, 1,82-96

57 Thompson, K. , Schofield, P. (2007). 'An investigation of the relationship between public transport performance and destination satisfaction'. Journal of Transport Geography. 15, 2, 136-144

58 Urry, J. (1992). 'The tourist gaze 'revisited''. American Behavioral Scientist. 36, 1, 172-186

59 Wilkins, H. (2010). 'Using importance-performance analysis to appreciate satisfaction in hotels'. Journal of Hospitality Marketing and Management. 19, 8, 866-888

60 World Travel and Tourism Council (WTTC) (2011). Travel & Tourism Economic Impact Report 2011. London, UK:WTTC

61 Yoon, Y. , Gursoy, D. , Chen, J.S. (2001). 'Validating a tourism development theory with structural equation modeling'. Tourism Management. 22, 4, 363-372

Index